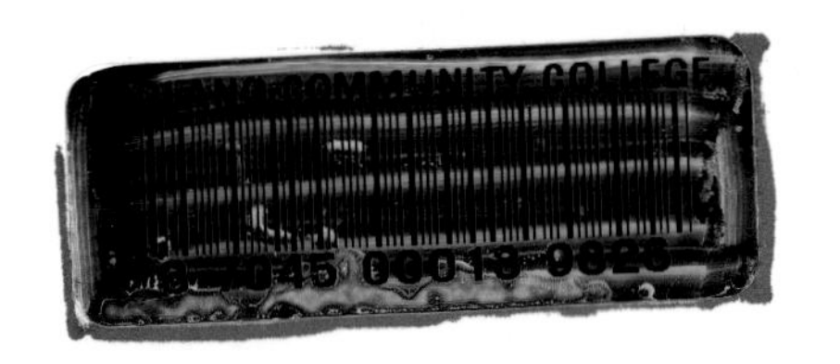

D0699707

HANDBOOK OF INTERACTIVE VIDEO

Steve and Beth Floyd
with
David Hon
Patrick McEntee
Kenneth G. O'Bryan
Michael Schwarz

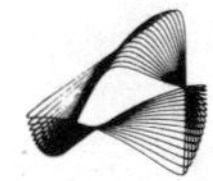

Knowledge Industry Publications, Inc.
White Plains, New York

Video Bookshelf

Handbook of Interactive Video

Library of Congress Cataloging in Publication Data
Main entry under title:

Handbook of interactive video.

(Video bookshelf)
Bibliography: p.
Includes index.
Contents: Introduction--Thinking interactively / by Steve Floyd -- Understanding interactive video: the technology / by Steve Floyd -- [etc.]
1. Video recordings--Production and direction. 2. Video recordings. I. Floyd, Steve. II. Floyd, Beth. III. Series.
PN1992.95.H36 1982 384 82-12690
ISBN 0-86729-019-6

Printed in the United States of America

10 9 8 7 6 5 4 3 2

Table of Contents

List of Tables, Figures and Appendixes

Illustrations

ACKNOWLEDGEMENTS

We wish to acknowledge the following individuals for their contributions to this book. Their review and suggestions are much appreciated: Al Bond, Texas Instruments; Jon Covington, Apple Computer, Inc.; Bob Howard, BCD Associates; Nicholas Iuppa and Tom Volotta, Bank of America; John McPherson, Sony Video Communications; and Jim Zinn, Pioneer Video, Inc.

Steve and Beth Floyd
July 1982

Introduction

by Steve Floyd

Trying to write objectively about a topic as exciting and dynamic as interactive video has been extremely challenging. It is no secret that interactive video has stimulated a great deal of interest, anticipation and confusion in the video community. Obviously, everyone is very excited about the implications of interactive technology. However, this enthusiasm has been dampened a bit by concerns about the very real obstacles to implementing this new technology.

Most of the current discussion surrounding interactive video has focused on hardware capabilities rather than design or planning considerations. Some proponents have described the new technology as a revolutionary step in communications, one that will lead to the inevitable decline of traditional linear video. On the other hand, critics, although less vocal, have characterized the new systems as expensive gadgetry which will fade as soon as the novelty wears off. This kind of debate generally raises more questions than it answers. In fact, some of the arguments are reminiscent of the "film versus video" debates that took place in the 1970s. Given this environment, it is not surprising that many of the critical issues affecting interactive technology have been obscured.

Using interactive video effectively requires more than simply knowing which buttons to push, or how to connect a microprocessor to a video player. Developing innovative programs requires at least three critical factors: commitment, time and money (probably in that order). Minimizing or ignoring the discipline, cost, risk and energy required to implement effective programs actually hinders the kind of thoughtful risk-taking that leads to successful programs. Therefore, this book attempts to shed some light on the critical factors that determine a program's effectiveness.

This *Handbook* was written to clarify many of the misconceptions surrounding interactive video. We have assembled a diverse group of voices from different disciplines to give the reader the benefit of a variety of perspectives. Although the individual authors may disagree on specific points, they share a common professional goal of improving the effectiveness and quality of video programming. With this goal in mind the book has been organized to achieve three objectives:

1) To increase the reader's awareness of the wide range of possible applications, as well as some of the specific requirements necessary to implement innovative interactive video programs.
2) To help the reader implement the techniques and principles presented in cost-effective applications.
3) To minimize the inherent risk of introducing new technology into an organization.

By organizing the book around these three basic objectives we have tried to make the information interesting and straightforward. The book is written for a broad audience of training and communication professionals. We have tried to keep the material free of professional jargon so that it can be used effectively as a resource by instructional designers, video professionals, trainers, educators and AV specialists. We also tried to show how all of these disciplines bring unique skills and perspectives which will blend easily with interactive techniques. The key is simply to remain open and flexible to fresh approaches.

We hope that by providing you with the prerequisites for interactive programming you will be better prepared to use it to improve your organization's communication system. Your ability to manage the development, production and implementation of these programs will ultimately determine whether interactive video will be a dramatic step into a new phase of video communications, or just another expensive gimmick. For this reason, we have structured the *Handbook* to reinforce the human factors of creative design and execution, rather than the specifics of selecting an individual hardware system. We hope that you are able to use the *Handbook* effectively in meeting the challenges of interactive programming.

1

Thinking Interactively

by Steve Floyd

When a personal computer is combined with a video player (tape or disc) the whole becomes much greater than the sum of the parts. Interactive video, the current marriage of these two technologies, is opening up bold new possibilities for personalizing and restructuring the ways we communicate.

For example, instead of passively watching a video program about landing a plane, a trainee can actively participate in an interactive video program in which he actually selects the next move. Or, rather than struggling with a maintenance manual, a mechanic can perform complex diagnostic procedures by responding to a series of program prompts, or directions, that branch him to the most probable corrective action. Recent developments in microprocessor technology have made training and communications applications like these seem almost commonplace. Almost.

Today, hardware is only part of the story. Although a great deal of attention currently focuses on the capabilities of interactive video systems, the real challenge is not which system to install but how to integrate and utilize these new technologies so that they have the greatest impact on, and benefit for, the user.

PROCESS ORIENTATION

Unfortunately, there is a general tendency to treat new hybrid technologies as though they were simply enhanced versions of the old technologies.

Approaching interactive video as if it were only a new feature of a popular product restricts its impact to a fraction of its potential. In order to comprehend how these new communication technologies can dramatically affect work habits and alter many of our daily routines, we need to approach technology as a process, not as a thing. By looking at technology as a process, instead of as equipment or hardware, we begin to see the importance of rethinking our orientation to computers and video.

Let's use the example of word processing technology to illustrate the importance of a process orientation. Word processing systems have made it desirable to centralize typing services so that a secretary can perform more administrative jobs. The transfer of this "mechanical" chore of typing to a separate department has redefined the secretary's job. As the secretary assumes greater operational responsibility, managers are free to focus on more critical issues. This, ideally, creates more satisfying work for a secretary and a more productive organization.

However, if a word processing system is treated as just another piece of hardware (say, as a sophisticated typewriter) rather than as a process, the equipment would probably remain at the secretary's desk and only one task would be affected—typing. The process approach enables reorganization of the workload, allows people to assume more responsibility and increase their effectiveness, and has a much greater impact on productivity.

We use this example for two reasons. First, it underlines how technology can dramatically change business procedures when the process approach is applied. Second, the tendency to think of word processors as just sophisticated typewriters, instead of as a process, is the same approach that most people take when they begin evaluating interactive video. People tend to focus only on the capabilities of the equipment and not on the potential impact of the process on their current methods of communication.

DEFINITION

Interactive Video

Keep this process orientation to technology in mind as we define interactive video. For the purposes of this book interactive video is defined as *any video program in which the sequence and selection of messages is determined by the user's response to the material.* Viewer participation in an interactive video program is a critical factor; this is not so for traditional linear video programs, in which the viewer remains passive. Viewer participation or involvement may take a number of different forms, such as: answering questions, manipulating a control during simulation or simply choosing which segment of material to view from a menu. As a result, within the same program a number of alternatives or paths are available to different users, and the route followed by each individual may vary significantly.

The critical difference between interactive and linear programs is simply that the rate, sequence and selection of information is a product of the user's active involvement with the program. This is true whether the program is highly structured with tight branching paths or loosely structured with the user guiding the program.

Linear Video

Traditional, linear video currently is used for most applications. Linear video generally does not require a response from the viewer. The program's structure, pace, selection of material and sequence is always the same. In a linear video program these decisions are determined before the program is produced and edited. As a result, the structure, sequence, pacing and content are the same for every viewer.

This is an important distinction to keep in mind: when you design an interactive lesson you are putting together material which may have a multitude of branches or paths available to the user. The particular path will depend upon how the user interacts with the material. Interactive programs, unlike traditional linear video programs, require a viewer to actively engage the material, as he or she works through a program.

LEVELS OF INTERACTION

Let's use an example to illustrate this distinction: developing a video lesson to train a salesforce on how to sell a specific piece of equipment. Let's start first with developing a linear video program. A linear video program depicts a sequence of events using a logical presentation of the material. To begin with, the linear video program would present the technical product information, such as the construction, controls, calibration and maintenance procedures. After presenting the product information it might depict a scenario in which a salesman holds a customer's attention, explains benefits, answers objections and finally closes the sale. Remember that all of these topics would be covered in the same sequence and at the same rate for each viewer or trainee.

Programmed Instruction

By contrast, to develop a fairly basic interactive lesson you might begin with a pretest covering product knowledge. The trainee would then proceed to the specific areas in which he needed additional product training.

For example, if a new salesman understood how to perform all of the computations for calibrating the equipment, but had never actually operated the equipment, he would be routed through all of the segments except the section on calibrating the test results. Another trainee might be an old hand at selling this type of equipment but unfamiliar with this particular model. As a result, he might skip all of the segments except the ones about new controls and maintenance. Both trainees would be branched according to their responses on the pretest. You could also include a post-test after each segment so that if the trainee failed to demonstrate proficiency he or she would be routed to the appropriate section to review the material, a process known as *branching*. The diagram for this simple type of branching is illustrated in Figure 1.1. The same logic is also applied to routing trainees through the second part of the course on sales techniques with each trainee branched according to how he or she responded to the questions posed throughout the course.

This example highlights some critical differences between linear and interactive video. However, the approach taken in the interactive lesson is really nothing new; it simply applies *programmed learning instruction* (PLI) techniques to modify a traditional linear video approach. The important point is that the lesson is now geared to each participant's needs.

There are at least two other levels of interactive design which go beyond basic programmed learning. For lack of a better title we'll refer to them as *programmed simulation* and *exploratory simulation*. Let's examine how we can apply each of these levels of design to the same training lesson we used in the earlier example.

Programmed Simulation

A programmed simulation uses many of the concepts employed in highly structured branched programs. The main difference is that the information is presented as a simulation of a real situation, with questions posed at the critical decision points, instead of being presented in units *about* each topic. The user's response to a question branches him to the appropriate action following his response.

For example, instead of presenting all of the information in separate segments with pretests and post-tests, the program might begin with a simulation, in this case of the salesman opening a sales call. Following the salesman's opening remarks to the customer the program would stop at a critical decision point, at which time the user would be asked what to do next. The program might ask: would you probe to uncover business needs, would you begin explaining benefits, would you treat the customer's

Figure 1.1 Flow Chart for Simple Branching

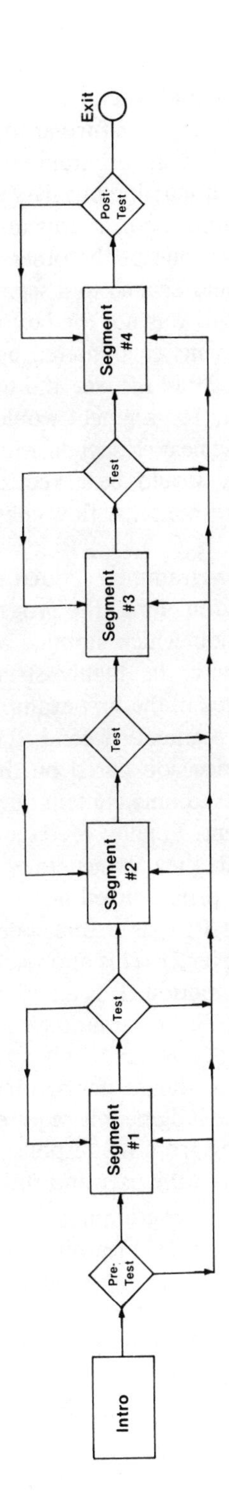

remarks as an objection or would you try to close the sale? In this example, let's say the best answer was to probe further to uncover business needs. If the trainee selected that answer the program would continue, showing the salesman and the customer discussing business needs and the action in the sales call would continue until the next critical decision point.

If the trainee had selected one of the other answers, say, "explaining benefits," the program would branch to a segment showing the salesman explaining benefits. Since this was not the best answer the narrator might highlight some of the problems encountered by the salesman during this phase, and explain how he could recover and use his probing skills to get the sales call back on track. The segment would continue along this path until it was time to make the next critical decision. Here, again, a question about what action to take would be asked and the learner would be branched according to the response. A flow chart for this type of program is illustrated in Figure 1.2.

Now let's contrast this programmed simulation with the simpler example of programmed instruction. First, the programmed simulation used a simulation of a real situation which stopped at critical decision points. Compare this approach with the highly structured, tightly organized branching sequences followed in the first example, when content information was presented in each segment. Second, the questions in the simulation required a judgment decision based on the action in the program, while the earlier example asked only content questions about the information covered in each segment. Finally, pretests and post-tests usually are not used in simulation, while they are generally used as important criteria to evaluate performance in programmed instruction.

The first example, basic PLI, is a direct descendent of the teaching machines (reviewed in Chapter 2). This approach is highly structured with a distinct hierarchy of information. It is an excellent example of didactic, or systematic, instruction. The programmed simulation example is not nearly as tightly structured; it does not rely strictly upon reinforcing the appropriate response. Rather, the learner becomes more intimately involved with the program because judgment is required; he sees a simulation of the probable action resulting from his response. Therefore the degree of involvement is higher between the user and the system.

Both levels of interaction, programmed instruction and programmed simulation, play an important role in learning. In the programmed instruction example the learner needs some ground floor technical information, and this information would be more difficult for the learner to process and comprehend if presented in a programmed simulation format. However, when the learner already has the technical knowledge, the programmed simulation is a powerful tool to begin applying that knowledge to realistic situations.

Figure 1.2 Branching Flow Chart for Programmed Simulation

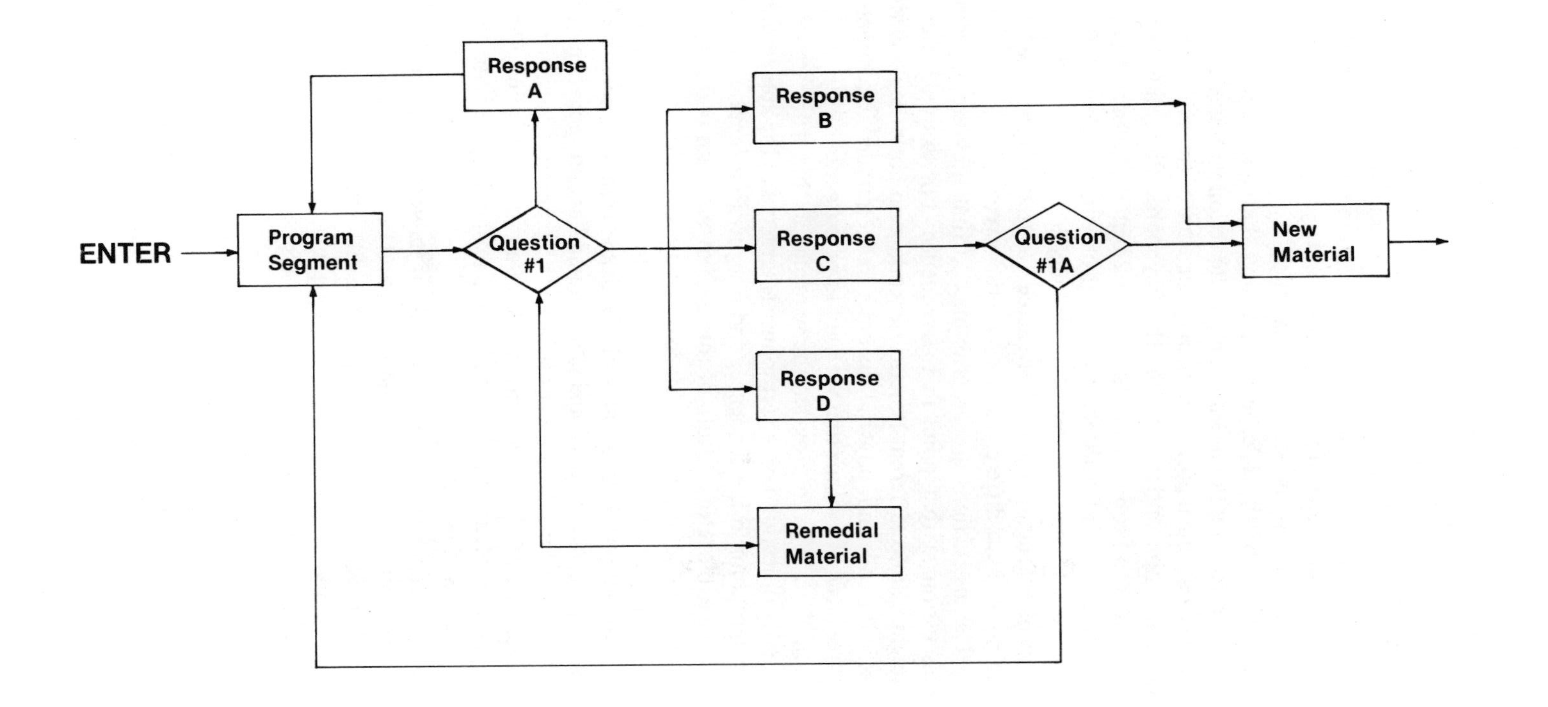

Exploratory Simulation

Now let's look at the third level of interaction, exploratory simulation. This level moves us even further away from the teaching machine concept to a learning environment resembling a video game. Let's use the same sales call as in the previous examples. The program will begin with a simulation of the salesperson opening the call. However, there are no specific decision points interrupting the action with questions. Instead, the trainee has much greater control of the flow of the call *at any time.* By hitting the appropriate key on the keyboard, the trainee can direct the salesman to probe for needs, explain a benefit, ignore an objection, tell a joke, acknowledge the customer's problem, or, close the call. Each of these commands correspond to keys which are coded with the appropriate type of instruction (in much the same way that different buttons on a video game produce different actions on the screen).

The instructions are programmed so that they are appropriate to the discussion at that point in the sequence. For example, the benefit key might call up different benefits depending on both the point the trainee is at in the call, and, how the customer has responded. Or, suppose there are three basic objections that we want to prepare the trainee for in this type of call. Here, we can program the sequence so that if the trainee presses the answer objection key, the response is targeted to the appropriate objection at that point in the sequence, and not just to a random objection.

The exploratory nature of this type of program makes it very complex from a design perspective, since there are so many options available to the user. However, notice how this program begins to resemble the same type of interaction we associate with electronic games. In fact, we have moved beyond the two dimensional plane of paddles and space ships to a very life-like simulation. Examples of this type of exploratory instruction have been developed at the Massachusetts Institute of Technology (MIT) and the American Heart Association. (Both projects are examined in detail in Chapter 8, Case Studies.)

After reviewing these three levels of interactive learning—programmed instruction, programmed simulation and exploratory simulation—you can see that each of these levels applies to different types of training and communication needs. As we move up the scale to exploratory simulation, hardware capabilities and software costs become important factors. More sophisticated branching at this level requires customized computer programming. The hardware is also more complex, more memory is needed, more video has to be produced, edited and stored in the system, and peripherals may be required. As a result, this type of program could cost two or three times as much to develop and produce as the simpler interactive program discussed in the first example. The point to emphasize is how far the third level of interaction takes us beyond the teaching machine concept, and the impact this type of interaction will have on communications.

IMPLICATIONS

In order to illustrate some of the implications of this new technology, it might be helpful to think back to before microprocessers were developed, when television was still black and white and home computers were straight out of Ray Bradbury.

Remember when someone got a new game like *Risk* or *Careers*? Everyone would sit around for thirty minutes trying to read the instructions printed on the inside of the box. Then we tried a practice game to test the rules. Finally, after endless rounds of debate (generally settled by referring back to the rules), each of us would begin to develop strategies based on our experiences with the game.

Let's contrast that with the action and excitement surrounding electronic games, in which the player learns to play by doing—by manipulating the controls. Next time you see an electronic game room, stop and read the instructions on the machines. There are usually four lines of simple pictorial operating instructions. The intricacies of the game, and any subtle strategies, are uncovered as the game is played.

Can you imagine trying to write a verbal explanation of the rules with appropriate actions for one of these games, or, for that matter, trying to write the instructions for a pinball machine? The explanation would probably resemble a driver's manual and all of us know how much fun they are to read.

The tremendous growth of computer games and video games is probably the best indication we have that something really significant is happening. Why are these new games so popular? Their novelty is certainly part of the answer, but there are a number of important differences that separate these games from their forerunners.

1) *They provide immediate feedback and gratification to the player.* It doesn't take hours to learn the rules or years to appreciate the subtle strategies of the game.
2) *The technology is transparent to the user.* In other words the player doesn't have to know how or why the game works. Instead, it's easy to play, consistent and dependable. The system doesn't interrupt the player's concentration; instead it sets the structure for the game.
3) *The games are not based on an oral, sequential language.* The

variables change so rapidly that a verbal explanation could not adequately express the nature of the game. Why spend an hour trying to describe something that takes three minutes to learn by doing?

These three concepts outline the fundamental differences between the traditional verbal-based or linear communication that preceded the introduction of the microprocessor, and the interactive communication that the microprocessor makes possible.

THE CRITICAL ELEMENT: DESIGN

At this point it is important to clarify a misconception that can easily arise: the most critical element in determining whether a program is effective is *design*—the planning and organization—not the delivery system. A poorly organized lesson is boring and counterproductive whether it's a live classroom presentation or an interactive video program. A video presentation which fails to highlight or demonstrate key points will fail to meet its objective whether it is linear or interactive. Just because a video program is interactive does not necessarily mean that it will lead to better learning or improved performance. Planning and organizing the material properly during the design stage are the critical factors that determine a program's effectiveness. (Appendix 1A, at the end of this chapter, presents some interactive video design tips to keep in mind.) This is one reason that it is often so difficult to revise an effective linear presentation to make it an effective interactive one.

Adapting Linear Programs

When we alter an existing linear program the best we can generally expect is basic programmed instruction with some branching. A linear program can be divided into segments with tests which loop the viewers back to review the program if they fail to demonstrate competency—but that's about it. In other words, it may not always be advisable to try to adapt existing linear programs for interactive segments. It can be effective but you may run into continuity and organization problems that were not apparent when the program followed a linear format. If there are legitimate reasons to justify this kind of adaptation, then be prepared—the results may be frustrating.

It's a bit like transferring a slide presentation to video. The information is still accurate but a slide transfer generally fails to take advantage of how video can enhance the message. You can compensate by using more shots with frequent edits, zooms, pans and special effects but this can be expensive. Unfortunately, without these extra touches an interesting slide–tape presentation often becomes a very flat piece of video. You may encounter similar problems in adapting a linear program to interactive.

WHEN INTERACTIVE IS THE WAY TO GO

Considering the additional expense and lead time required to design an effective interactive program it is important to focus on areas in which this type of approach both can be cost-effective and meet a unique need. The best situations to look for applications that meet these two criteria are when trainee salaries are extremely high or when the actual equipment is so expensive or dangerous that expensive simulations can be justified.

Training Applications

Let's begin with the first group for which salaries are extremely high and downtime for training can be prohibitively expensive. This is true for doctors, lawyers, pilots, computer programmers, executives and salespeople. The cost of not training these professionals to meet the subtleties of their jobs may far exceed any short term costs for training. Unfortunately, it is difficult, if not impossible, to measure the real cost of lost opportunities, financial planning errors, programming errors or legal oversights. Success in these jobs requires a combination of technical knowledge and professional judgment. Interactive training which reduces a pilot's or doctor's downtime has measurable benefits with direct return on an organization's investment. If you can reduce a six week training period for pilots to four or five weeks you can justify a sizable investment in interactive training. Similarly, if you can reduce a three-day sales training course to one day, you can realize tremendous savings.

Let's look at another area holding great potential for interactive training: high technology areas where the equipment is so expensive or potential errors so dangerous that on the job training needs to be supplemented by simulation. The most obvious examples are piloting an airplane, navigating an oil tanker, performing surgery, operating a nuclear reactor or diagnosing a problem in a computer system. In these examples, a programmed simulation could serve a very unique and pressing training need safely and relatively inexpensively. As a result, the investment for interactive video is justified because it provides a unique solution.

Consumer Applications

The other major area in which interactive video holds great promise is the consumer market. This category can be broken down into three groups:

learning, marketing and entertainment. Learning activities can be based on effective training techniques used in industry but they need to have broad consumer appeal, such as learning to drive a car, play tennis, bake lasagne or compute taxes. Increased leisure time and the growth of cable TV are heightening the demand for consumer applications such as these.

Another example of consumer programming is the use of random access video to demonstrate products or to answer typical questions about a product. This type of programming helps to stimulate sales by giving the consumer more in-depth information than he would generally receive from a sales clerk. General Motors developed this type of programming on video discs for its 6000 Chevrolet dealerships around the country.[1] In a related application, Handy Dan hardware stores nationwide have used interactive video effectively to demonstrate woodworking techniques in their stores.

Finally, the entertainment applications of interactive video may have the greatest impact on the consumer market. Video games manufactured by Atari, Mattel and other are already becoming a fixture in many homes. Combining the unique features of video games with actual video footage, so that the figures are not simple computer-generated graphics, creates a whole new dimension of video game.

Instead of just passively watching the chase scene from "The French Connection," imagine how exciting it would be to control one of the cars. Merging these two entertainment technologies—video games and traditional film—may open up an entirely new home entertainment medium. (However, one of the biggest obstacles to this use of interactive video could be the ensuing agrument over who owns the distribution rights for this new form of entertainment.)

As more entertainment applications develop, interactive video systems could very well become the "electric trains" of the future, as families begin to experiment and build their own systems.

These examples of consumer applications illustrate how both business and the home will change to meet the growing demands for information. As these marketing techniques become more successful in store locations, businesses will begin making interactive marketing programs available to consumers at home. Cable television companies are already looking at ways to make these interactive capabilities accessible to consumers through subscription services.

---

1. Efrem Sigel, Mark Schubin, Paul F. Merrill, et al. *Video Discs: The Technology, the Applications and the Future.* White Plains, NY: Knowledge Industry Publications, Inc., 1980.

SUMMARY

With these applications and design concepts in mind, we will look at ways to use interactive video to redirect traditional approaches to training and communication. We will look at the critical factors you must consider at each step of design, development, production and evaluation.

As you read the other chapters we hope that you will challenge yourself. Think about unique applications for your organization, and how you can use these techniques to actively involve the learner. Remember, we are following a process approach to refine and shape interactive video techniques.

APPENDIX 1A: INTERACTIVE VIDEO DESIGN TIPS

1) Think of how the typical user will perceive and react to the material. Make sure that he or she will be able to follow the logic. The material should not only be technically accurate but it should appeal to or stimulate the user as well.

2) Avoid negative or "punishing" answers to a user's response (a joke about a wrong answer may be taken the wrong way). Encourage or acknowledge the user positively, and correct any misconceptions as soon as possible.

3) Use the same techniques that work well in linear presentations:
    a) Motivate viewers by showing how they can use the material —why it is important and what the objectives are.
    b) Give an overview of the structure.
    c) Define major points and any unique abbreviations or acronyms.
    d) Use examples to support and highlight definitions.
    e) Process examples to clarify and reinforce key points.
    f) Tie the material together in a sub-summary before moving to the next major point or stopping for a test.

4) Make presentations visual and aesthetically pleasing. Avoid lengthy text displays or talking head explanations. Use symbols, color codes, numbers, titles and format consistently to reinforce continuity and to reduce the possibility for confusion.

5) Avoid true-false questions.

6) Apply "form follows function" to the design so that you always have a reason for an action. This gives the structure a logical direction and eliminates unnecessary branching or testing.

7) Make the technology as transparent as possible to the user—a standard already set by video games. Try to involve the user without using cumbersome peripherals, unnecessary testing, or long delays during a search. In other words, don't let the technology get in the way.

8) There is a definite learning curve effect in developing interactive programming. Early efforts may require two or three times the lead time and budget of later programs: so plan accordingly and don't become frustrated.

2

Understanding Interactive Video Technology

by Steve Floyd

Although recent developments in microprocessor technology have made the merger between random access video and computers possible, the evolution of interactive techniques has actually taken place over several decades. In order to grasp some of the implications of this new technology let's return to a time before microprocessors were invented, when television was a laboratory experiment and personal computers were straight out of Buck Rogers. It is surprising to find that many of the concepts we believe are revolutionary can be traced back to experiments in the 1920s.

BRIEF HISTORY OF INTERACTIVE TECHNOLOGY

Teaching Machines

The first teaching machine is generally credited to Sidney Pressey, a professor at Ohio State University. During the early 1920s, Pressey built a teaching apparatus that resembled a simplified typewriter. A multiple choice question was presented to the student in a window box. If the student pressed the key corresponding to the correct answer, the next question rotated into view. However, if the student pressed the wrong key, the machine recorded the incorrect answer and the student had to try again before advancing to the next question. Unfortunately, Pressey's attempt to revolutionize education through individualized learning and immediate feedback did not reform the educational establishment.

The next significant development in interactive technology came as a result of B.F. Skinner's research in the 1950s. Skinner and his group of researchers are credited with originating programmed instruction. The teaching machines they developed presented discrete units of information followed by fill-in-the-blank questions. However, within a few years the novelty of these machines wore off and gave way to linear programmed workbooks.

Dr. Norman Crowder introduced a major educational innovation by combining Skinner's and Pressey's work to create branched programmed tests. He designed multiple choice questions so that each answer routed the respondent to a different path. When the learner responded correctly, the answers were acknowledged and new information was presented. If the answer was incorrect, the response was followed by a different explanation. The student was then presented with a new question covering the old material. This programmed instruction approach was the forerunner of computer assisted instruction (CAI).

Computer Assisted Instruction (CAI)

Three IBM researchers, William Uttal, Nancy Anderson and Gustave Rath, gave birth to CAI in the late 1950s. Their experiment linked an IBM 650 computer to an electronic typewriter, so that elementary students could use the keyboard to learn binary math. The language developed in writing the program led to the IBM Coursewriter I authoring language. This type of tutorial approach combined with branched programmed instruction paved the way for a decade of educational innovations during the 1960s.

Two very comprehensive and successful CAI demonstration projects illustrate the range of experiments during this period: Programmed Logic for Automatic Teaching Operations (PLATO) and Time-shared Interactive Controlled Informational Television (TICCIT). PLATO was developed at the University of Illinois by a team of educators and engineers. This time-share system was designed so that hundreds of users at remote terminals could access CAI courses stored in a central computer. Subscribers, including the government, education and industry, used the telephone system to access materials covering a wide range of levels and topics. The TICCIT project was originally developed by the Mitre Corp. (McLean, VA) and Brigham Young University (UT). The system used a medium-sized computer and video technology to deliver highly innovative instruction to a number of users simultaneously. This concept was based on a medium-scale time-share system.

Video Discs

The CAI experiments in the 1960s and 1970s led to the current merger between video and the microcomputer. In 1977 McGraw-Hill funded an interactive video disc on biology developed by WICAT, Inc. (World Institute for Computer Assisted Teaching). This video disc project is generally recognized as the first interactive disc developed for individualized learning. In recent years MIT, the University of Nebraska, Utah State University and Brigham Young University have pioneered a number of innovative interactive video projects using video discs and microcomputers.

Over the last 50 years, interactive learning techniques evolved parallel to the hardware. Pressey's experiments with teaching machines took place at about the same time that television was still a laboratory experiment. Computers and television became major factors in our society while Skinner and Crowder developed programmed instruction. The rapid development of microprocessor technology over the last ten years has made sophisticated television equipment more portable, while making powerful microcomputers a reality. The same technology that made slide rules obsolete is now touching almost every aspect of our lives. As a result, we now have communications technology which has almost limitless possibilities for expanding our capability to solve problems.

COMPONENTS OF AN INTERACTIVE SYSTEM

With this brief "revisionist" tour of interactive technology in mind, let's look at the basic parts of an interactive system to see how the technology works.

Although interactive systems vary, they all include four basic pieces of hardware: a video player (either tape or disc); an interface; a computer (or microprocesser); and a television monitor or receiver. This configuration is illustrated in Figure 2.1. One additional element is the authoring system, or software, which enables you to give the computer commands. Since the authoring system is generally contained in a floppy disc, a disc drive unit is commonly used with the microcomputer.

Another requirement, needed for video tape players, is the random access capability of a solenoid driven player. (The solenoid allows a video player to accept commands from a computer—a feature lacking in many of the older video tape players.) If you plan to use a video disc player inter-

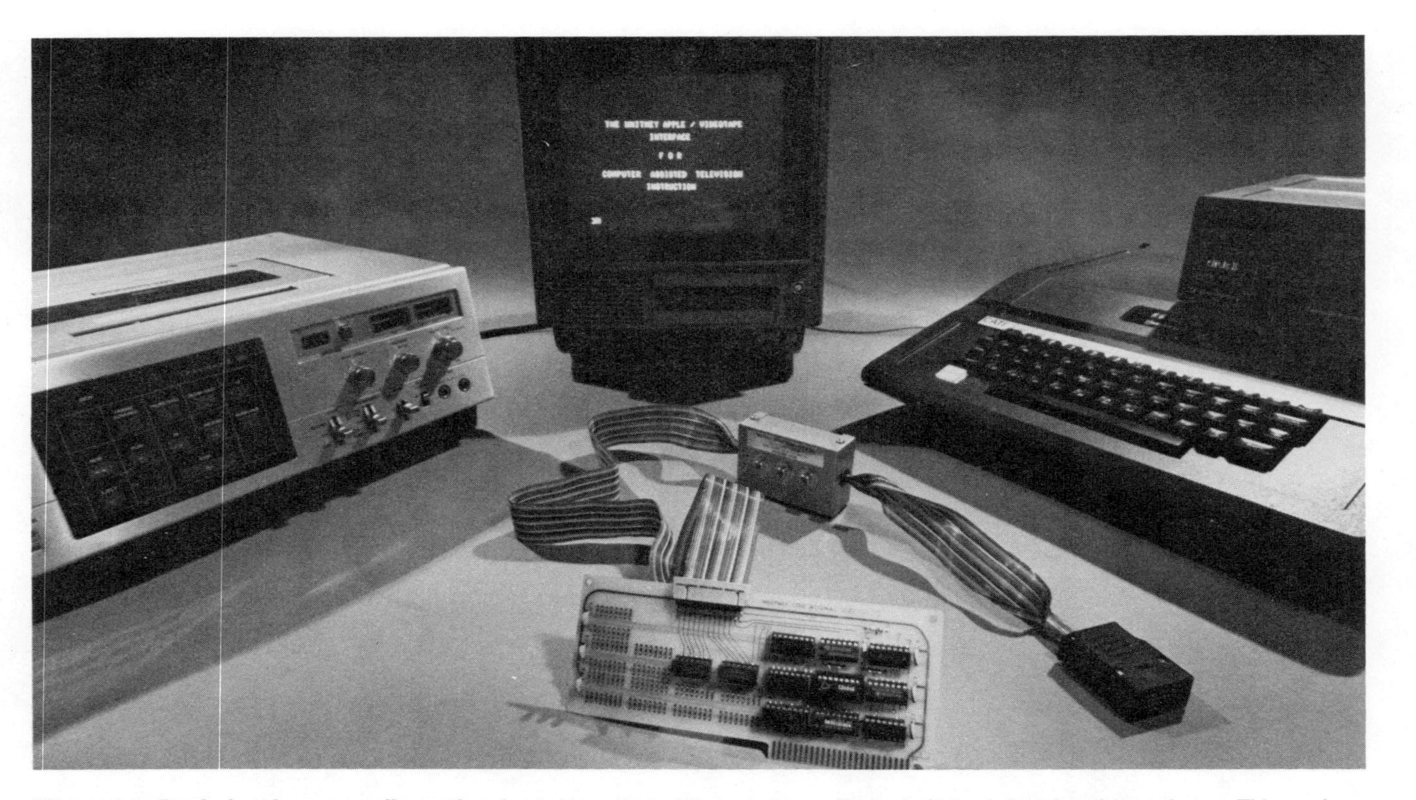

Figure 2.1 Basic hardware configuration for interactive video systems. From left to right, the video player, TV receiver, microcomputer, and in the foreground, the interface unit. Courtesy Whitney Educational Services, Inc.

actively you must use one of the laser optical systems listed in Chapter 3. The industrial models produced by Pioneer, Sony and Thomson-CSF are also equipped with an internal, programmable microprocessor. As a result, these laser optical video disc systems can be used interactively simply by programming the internal microprocessor. However, the internal microprocessor, with its limited capacity, can only accommodate simple forms of branching. For more sophisticated branching techniques, the video disc player (like the video tape player) must be connected to a microcomputer.

Interface

Let's examine each of these four components beginning with the interface. An interface is simply the component which allows a computer and another piece of equipment (e.g., another computer, a word processor, or here, a video player) to be linked together. The computer-video interface enables the computer to communicate directly with the video player control circuit. In most systems this interface is simply a printed circuit card which fits into a peripheral connector in the computer (see Figure 2.2). Other systems use a separate component with a standard RS-232 interface so that it can be easily connected to the computer. The interface performs three critical functions:

1) Transmits commands for controlling the video player;
2) Processes signals from both the computer and the video player relaying the appropriate signal for display on the television screen;
3) Counts control track pulses or video frames depending on the system. This function enables the interface to locate, synchronize and display the correct material.

The interface will work with both solenoid driven (random access) video tape players or industrial video disc players. Solenoid driven players are used because the control keys on the front of the player activate the control circuits which energize the solenoids, causing the appropriate mechanical action.

Video Tape Players

By reviewing how a video tape player functions, it is possible to see how the interface and player work together. A video tape player records a succession of signals on tracks along a linear tape path. These signals are recorded at 30 frames per second (or one frame every 1/30 of a second).

Figure 2.2 The Apple Computer uses a printed circuit card interface which plugs into computer expansion slots. Courtesy Apple Computer Inc.

When an interface is connected to a video tape player, a control track signal is transmitted to the computer. A control track is simply an electronic pulse which is recorded on a separate track of the video tape every 1/60 of a second. The control track is also used to synchronize two tape machines for editing. These pulses enable specific addresses or locations to be found on the tape. Several interactive systems provide an optional time code module which records a time code on an unused audio track. (The time code is a recorded imprint on the video tape, measuring hours, minutes, seconds and frames.) The time code is then read by the computer to locate addresses on the tape. This time-code technique is more accurate than counting control track pulses, which are not always frame accurate.

Since video tape often slips with constant shuttling between rewind and fast forward, slight errors of several frames can build up as a program shuttles between segments. Most systems using control track counting provide a tolerance of four to seven frames accuracy when branching to another segment. This simply means that a tape might be off approximately 1/5 of a second—tolerance acceptable for most user applications. Arranging the segments in sequence in the order that they will most probably be chosen reduces the amount of tape shuttling. It also increases frame accuracy and decreases the amount of time needed for shuttling between segments.

Laser Optical Video Disc

It is useful to contrast the operation of a laser optical video disc player with a video tape player. Instead of recording the signal on tracks along a linear tape path, a video disc records each frame of video as one rotation on the disc. Therefore each frame number corresponds to a frame on the disc. The Pioneer, Philips and Sony discs spin at 1800 rpm on normal playback speed. Since the disc has a capacity of 54,000 frames per side the normal playback time is 30 minutes.

Playback time can be extended to 60 minutes by switching from CAV (Constant Angular Velocity) to CLV (Constant Linear Velocity), but the CLV mode sacrifices the interactive programming capabilities of the video disc. A CAV disc rotates at 1800 rpm while the CLV disc rotates at a speed ranging from 1800 rpm at the inside of the disc, to 700 rpm on the outside. Decreasing the rotation speed extends the normal playback time but, in the CLV mode, only normal play and scan operations are possible. With a CAV disc, playback at variable speeds, frame number display and programmable operations are also possible. As a result, the CAV mode is the only alternative for designing interactive discs. The RCA and JVC capacitance discs also rotate at lower speeds to achieve a longer playing time of 60 minutes per side. For our purposes, however, we will refer only to the capabilities of optical laser discs in the CAV mode.

Storing the signal on concentric frames instead of on a linear tape path gives the disc its most important feature for interactive video: rapid, frame accurate access to any segment on the disc (maximum search time is three to five seconds). Contrast that capability with the video tape player. The video signal is recorded linearly along a section of tape. Therefore, the access time to branch the viewer to another segment, or to search for an item on the program menu, is dependent upon two factors: the distance the tape has to travel and the mechanical speed of the machine. Although the fast forward and rewind functions for VHS and Beta format players are fast compared to the ¾-inch format, they may take a number of seconds to locate the right segment. Remember, a video tape player counts the pulses on the control track or time code to locate the right frame before it even begins to lock into play. As a result, search time can be agonizingly slow if the segments are not located near one another.

As mentioned earlier, search time can be reduced by structuring the tape so that the most probable branches of sequences are batched together in close proximity. (Chapter 4 discusses this aspect of interactive design more fully.) This technique reduces the distance, and time, necessary for shuttling tape. Other techniques can be used to disguise the search time, such as carefully sequencing the material or presenting information from the computer while the tape player is searching. Regardless of the techniques used to maintain the viewer's attention, search time and accuracy are important advantages of the video disc.

Tape or Disc

In order to present a balanced picture of both delivery systems (tape and disc) let's review some of the advantages and disadvantages of both systems. The major advantages of video disc are:

1) Rapid access to any frame.
2) Unlimited still frame capability, with no wear on the disc.
3) Precise frame accuracy in accessing any frame.
4) Scanning of material at 100 times playback.
5) Advance frame-by-frame or 1/5 playback speed.
6) Extremely low duplication cost for large numbers of copies.
7) Internal programmable microprocessor for basic interactive programs.
8) Greater storage density.

Video tape systems have the following advantages:

1) Record capability.
2) Duplication locally; no four to six week delays waiting for a disc master.
3) No mastering charge ($1500 to $2500+ per disc program).
4) Often a network already exists so that video equipment does not have to be purchased (providing that the existing players are solenoid). Additional expenditures are for the microprocessor (or microcomputer) and an interface.

As this comparison indicates, an effective interactive video system can be implemented with either tape or disc. The choice of a tape or disc system really depends on your needs. For example, if search time, frame accurate access and large numbers of copies are critical factors in your planning, then disc may be the most logical choice. However, if you already have a successful video tape network in place, and if frame accurate access and search time are lower priorities, you might select a video tape system. Since unique applications can slant a recommendation in either direction, it is safest to evaluate each application separately. The long range planning implications and overall costs of either alternative require careful study before any system is selected. The safest alternative is usually to plan any investment in phases, so that the organization sees a measurable return as the system expands.

Television Receiver or Monitor

The most common component in the system is the television monitor or receiver. The television receiver performs two important functions:

- It displays any visual information transmitted from the video player or the computer.
- It also receives and amplifies any audio signals it receives from the video player.

Receiving and amplifying *audio* signals distinguishes the receiver from a CRT or computer display, because a CRT only displays *visual* information. Both of these functions make the receiver the output vehicle for any visual or audio information transmitted from the system. However, since the receiver cannot store information, two other peripheral devices are frequently used: hard copy printers and disc drive units. These devices may be used to store information about the learner's performance, such as number of correct answers, time using the equipment or overall rating.

The Microcomputer

The final component in the interactive system is the microcomputer or microprocessor. As mentioned previously, some video disc systems contain built-in microprocessors, capable of performing simple branching functions. For more sophisticated branching, the increased capacity of an external microcomputer would be required. For the purposes of this book, we will define a microcomputer as a system of circuits that perform specific functions in a prearranged order according to the commands received. A microcomputer has four basic functions: input, memory, processing and output. In order to give a better idea of how a microcomputer works, we will review each of these functions in turn.

Input is simply how the user communicates with the computer. The most common form of input is a keyboard, or, in some video disc systems, a keypad. Some systems also use touch-sensitive television screens which route responses back to the computer through the television receiver (see Figures 2.3 and 2.4). The second function, *memory,* stores information that the computer needs to make decisions. This information might be in the form of rules, or step-by-step procedures. The *processing* function decides how to perform the commands, based on user input and the memory already stored in the computer. Finally, the *output* function causes something to happen as a result of processing the input. This action may take the form of displaying information or it might be controlling the operation of the video player.

That's all a computer really does: it receives informations from input, processes it, makes a decision to act and causes something to happen based on the input and memory. These four simple functions form the heart of the interactive video system. Without the computer or microprocessor, we have a linear system that plays back video material exactly as it was recorded, with no variations. By adding the microcomputer we have created an entirely new hybrid system—one that incorporates the capabilities of each component so that the total system is really much greater than the sum of the individual parts.

Authoring Systems

The previous sections discussed the individual pieces of hardware that make up an interactive system. The one element missing is the software, or authoring system, that makes the hardware work together. (The software, or authoring system, referred to here should not be confused with the final interactive video program itself. The authoring system is, essentially, the intermediary between the hardware and the user.) In order to make a program interactive you need to enter commands into the computer. This is accomplished by using an authoring system. An authoring system enables the message designer to enter all of the sequences and events in a step-by-step procedure, according to a lesson plan. This is done by using a series of prompts, or questions, to convert a lesson into an interactive program, complete with computer generated test, video segments, questions and branching paths. The author's instructions are translated into machine language commands so that the appropriate segment or information follows each response. As a result, nonprogrammers can enter lessons into a system without knowing computer language or syntax. Authoring systems vary from the basic instant prompts of the Sony Recorder to floppy discs containing easy-to-follow directions for authoring.

Although you do not have to understand computer programming terminology, you do have to remember that the computer performs exactly the command you enter through the authoring system. As a result, you will need to familiarize yourself with the logic of the authoring system you are using. For example, if you enter a command to show a page of text and fail to enter the length of time the text is to remain on screen, when the learner runs the program, the page of text will flash on screen and then go immediately to the next segment in the sequence. The point is that although the authoring system eliminates the need to create a special computer program for each lesson, it still requires the designer to think in terms of what the instructions mean to the computer.

THE OPERATION

The next step is to review what actually happens when a typical viewer sits down to work with the system (see Figure 2.5). First, the viewer makes some sort of entry into the system (input). This can take several forms, as discussed earlier, but generally, information is entered by using a keyboard. Pressing the keys sends a signal to the microprocessor. Depending upon how the programmer designed the system, the microprocessor will either display computer text or search for a video segment to display. In this case, let's assume the program calls for displaying a page of computer text which is a menu of the possible video segments available to the viewer. The computer relays this information directly to the television receiver which then displays the menu. The video player will not display any of the segments until it receives new input.

Next, the viewer selects one of the video segments and presses the appropriate key on the keyboard. On the basis of this input the computer issues a command to the video player to search and play the appropriate segment. As soon as this segment is located, the video player sends the audio and video signals to the television receiver where the video signal is displayed and the audio signal is amplified.

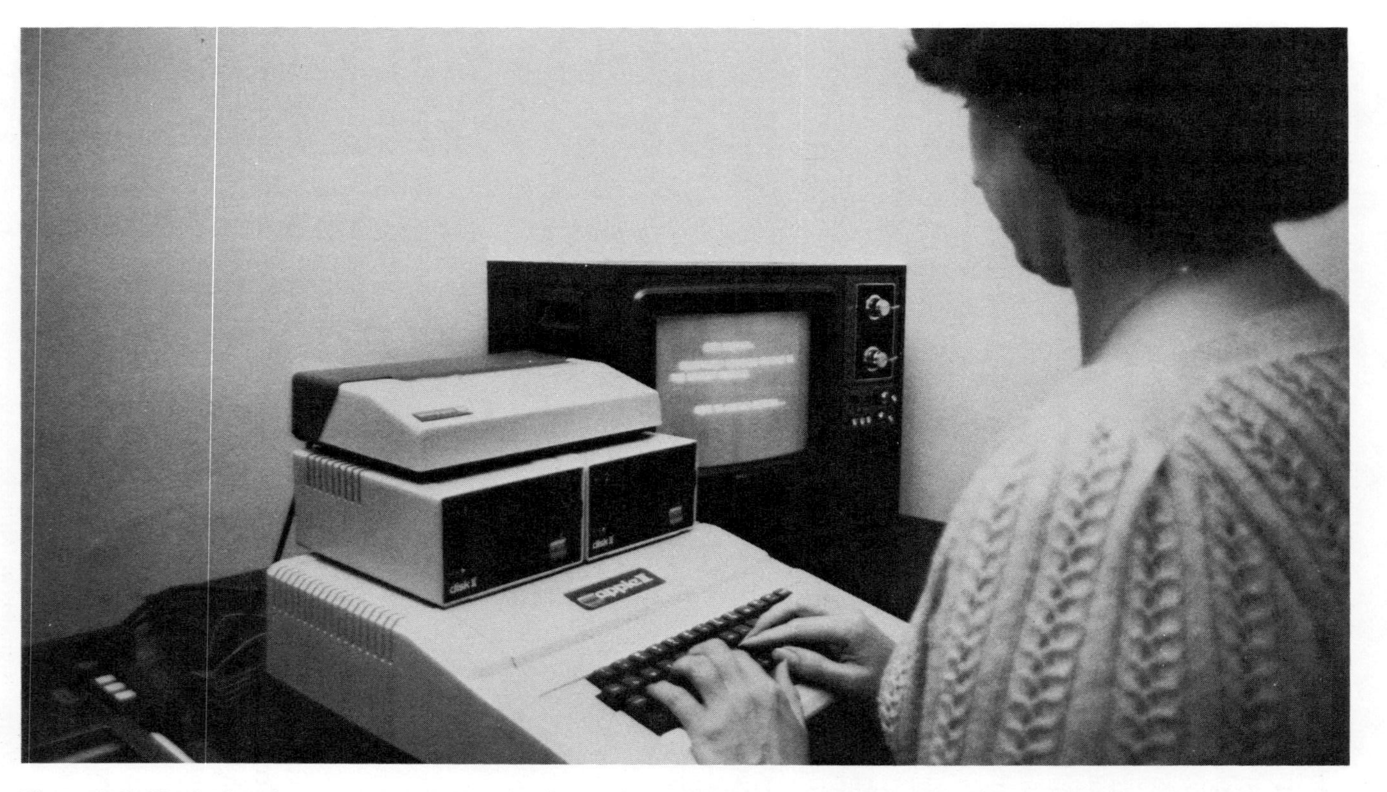

Figure 2.3 Student enters a response to question, using a keyboard. Courtesy Minneapolis Medical Research Foundation, Inc.

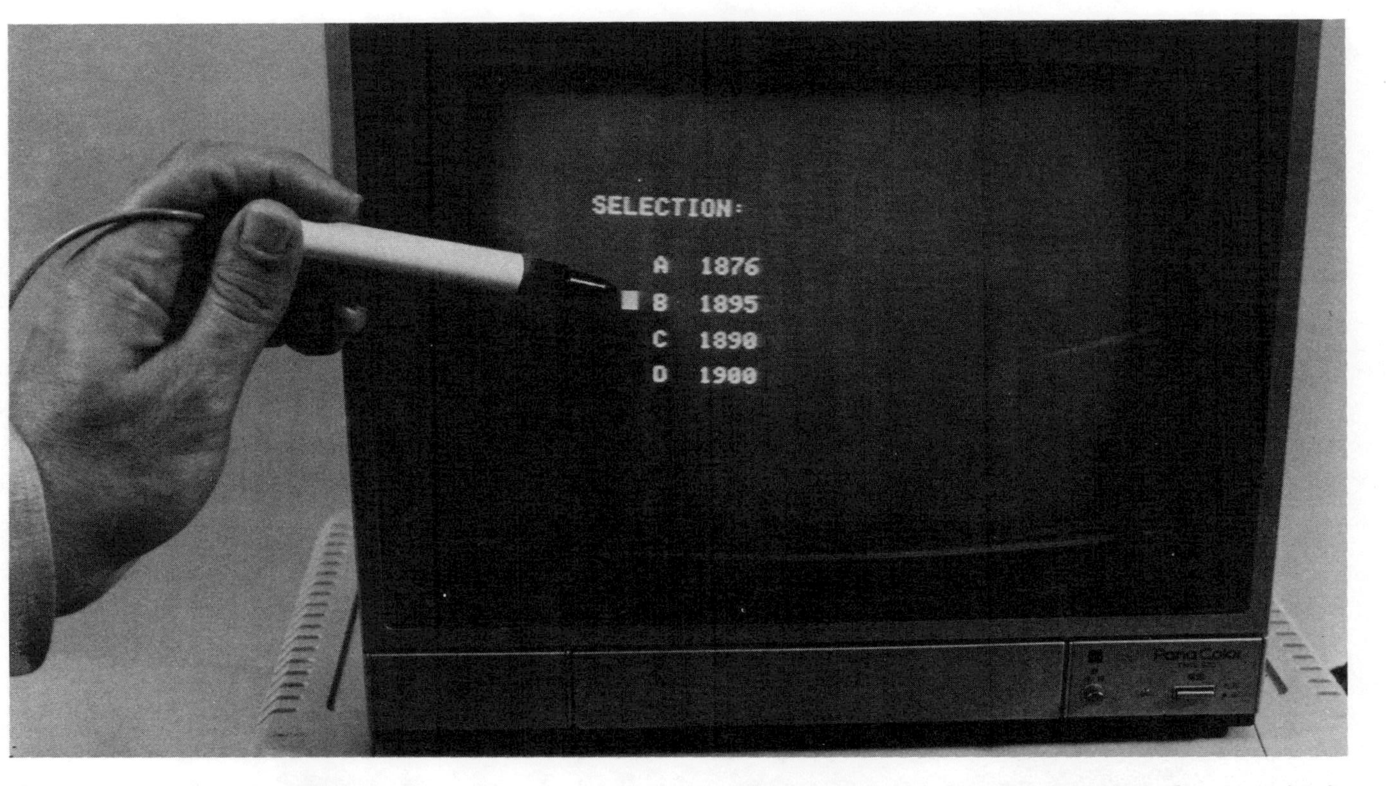

Figure 2.4 By use of a light pen, a user can communicate with the system without a keyboard. Courtesy Apple Computer, Inc.

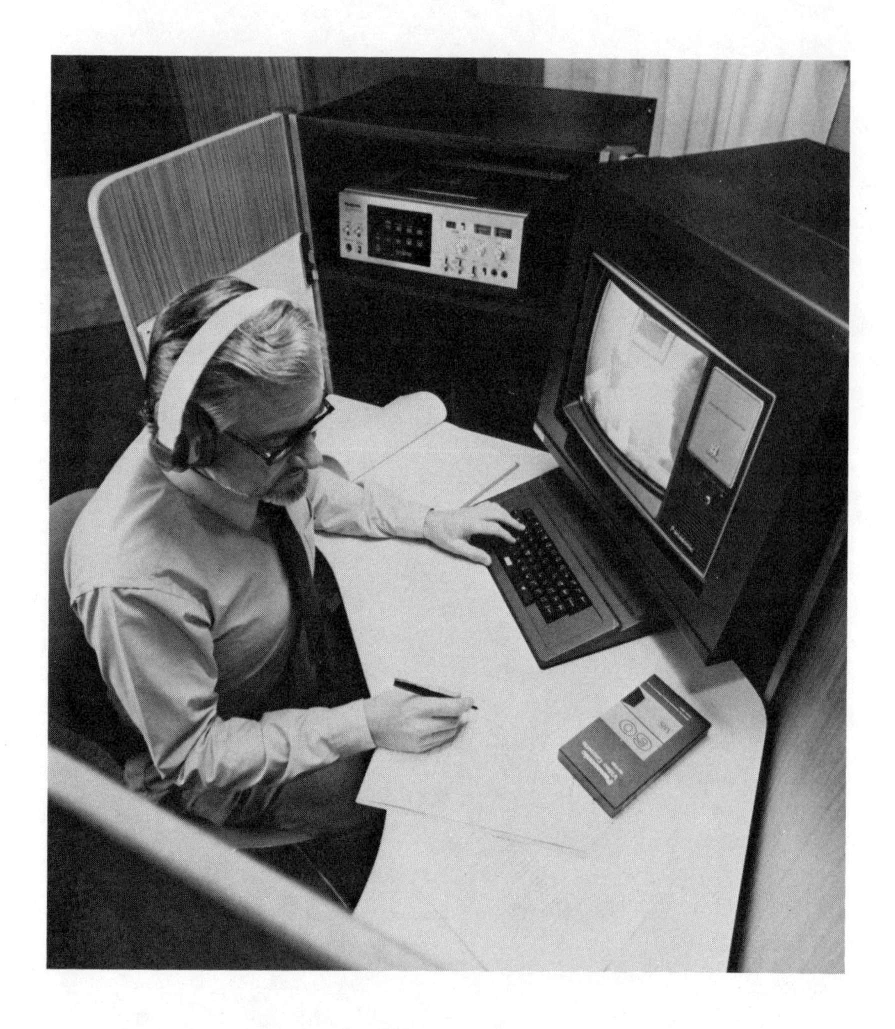

Figure 2.5 Often the interactive video learning process can take place inside a study carrel system, such as the Synsor LEM carrel shown here. Courtesy Synsor Corp.

When the video segment is completed, the video player transmits a signal which initiates display of a page of computer text that asks a multiple choice question. At this point the computer cannot call up another segment or display any more text until the viewer responds by pressing any of several keys "A," "B," "C," "D," etc., which are programmed by the author to give new input to the microprocessor. (The author can also program the computer so that if no input is entered by the viewer within a specified time of displaying the question, then a clarifying question is displayed, or the earlier video segment replayed.)

Let's assume that the viewer pressed one of four keys ("A," "B," "C," "D") which were programmed for a response. Regardless of the key selected, the viewer has entered new input. In this example, say that key "A" called for a new video segment and key "B" required a replay of the same video segment. The "C" key simply displayed computer text and asked another question, while the "D" key called for another video segment, different than "A." The computer combines the input (the key pressed) and the memory (the instructions already programmed into the memory) to determine what to do next. In this case, the alternatives are to display one of three video segments or a page of computer text.

SUMMARY

This cycle of input into the system, followed by branching the viewer to the appropriate material, followed by another decision point requiring new input from the viewer, continues until the viewer has either successfully completed the program, or failed.

This example also demonstrates the "feedback and control" characteristics of an interactive system. The cycle requires ongoing feedback and participation by the viewer. If the viewer fails to respond or provide input into the system, he or she cannot advance through the program. This requirement of feedback, or input, into the system is the critical element that distinguishes interactive technology from traditional, linear video. All four components, the player, the receiver, the interface and the computer, work together to complete the interactive cycle. The authoring system provides the sequence of commands or instructions for each component's actions. Although the computer (microprocessor or microcomputer) is the heart of the interactive system, it is equally dependent upon the other components to complete the cycle.

Ultimately, the author (i.e., the software designer), not the manufacturer, is responsible for using the full range of capabilities that the technology makes possible. As a result, the challenge is much greater for the designer since more variables are involved than the creators of teaching machines or television sets could have possibly imagined. Fortunately, the limits to the technology are determined mainly by our ability to create new software applications. As this review of technology indicates, the basic hardware configuration and mechanics of an interactive video system are relatively simple. Do not become overly concerned about whether the computer is a peripheral device for the video player, or vice versa. Focus on the process, not the hardware, since the real difficulty is in realizing the potential of interactive video to improve communication and learning.

3

Interactive Video Equipment

by Steve and Beth Floyd

Interactive video equipment is undergoing dramatic change and innovation as manufacturers compete to develop hardware that meets the demands of the marketplace. Trying to match equipment to these demands is a difficult and risky undertaking, requiring large amounts of capital investment. As a result of these rapid changes, the hardware will continue to evolve until a system is introduced which alters the complexion of the industry—as profoundly, perhaps, as Sony's introduction of the ¾-inch U-matic format affected portable video recording in the early 1970s.

If a similar innovation sweeps the interactive video market, components and systems will be designed to be compatible with a single system. However, when and if this development takes place is still pure speculation. Furthermore, just because this possibility exists, does not mean that you should postpone implementing an interactive video network. It does mean that you should select a system that is both flexible and expandable, so that new components with greater capabilities can be introduced without making the system totally obsolete. This same principle of planned implementation should also be applied to expanding or adapting an existing linear video network to make it interactive.

TURNKEY VS. COMPONENT SYSTEMS

Keeping these points in mind, there are two basic approaches to selecting an interactive system: turnkey or component. The turnkey approach refers to purchase of the entire system—the interface, microcomputer, video player, authoring system, etc.,—from one source. This approach has several advantages:

1) It ensures compatibility among components.
2) Service requests go to one source.
3) Since the components are purchased as a system, instead of individually, there is some opportunity for savings.

The other alternative is to select each component in the interactive system independently, rather than as an entire package. This alternative also has some advantages:

1) It is possible to select the best component for specific needs.
2) The system can be built in phases and expanded as it succeeds.
3) Different components can be experimented with to create a customized system.

Both approaches have some drawbacks as well. For example, selecting each component individually runs the risk that certain components may not be compatible with others, unless clearly specified by the manufacturer. Also, this approach can be time-consuming since it requires more involvement and evaluation. Often, it may be reinventing the wheel since many of these steps have already been performed by the manufacturers.

The turnkey approach also presents some disadvantages. The system may not be flexible enough to meet all future needs. Also, since a turnkey system was probably designed to meet the needs of the majority of users, individual requirements or unique applications may be compromised. Finally, turnkey systems are usually designed with specific pieces of equipment in mind. These specific units may not be compatible with equipment already installed in a network.

Obviously, both alternatives have critical factors that should be evaluated carefully according to an organization's operational needs, before committing to any investment.

EVALUATING AN INTERACTIVE VIDEO SYSTEM

Regardless of which approach is used, there are four major areas that should be addressed in evaluating a system. These are dependability, service, features and costs.

1) *Dependability.* Does the manufacturer have a proven track record? Can it deliver on-time within budget? Will replacement parts be a problem? Is the manufacturer likely to stay in business?
2) *Service.* Is local service available? How will the system be maintained? What kinds of warranties are available? Has the service record been good?

3) *Features.* What capabilities are essential? Can the system be upgraded easily? Is the system flexible? What kinds of peripheral devices (light pen, printer, etc.) are available? Is the system compatible with other types of hardware, or does it have to come from a single manufacturer?
4) *Costs.* What is the total cost of the entire system (programmer, authoring software, support materials, player, etc.)? Are additional capabilities expensive to add? Is there an additional charge for a service agreement? Is there a licensing fee or lease arrangement?

Depending on the applications and organizational needs, other factors may also have to be considered. Furthermore, each user may weigh these factors differently. For example, someone in a small organization might be more concerned with service and support than someone in a major corporation where there is a wider range of resources available. Large manufacturers with a sizable marketing force may or may not be as reliable and dependable as a small manufacturer just entering the field who really needs the account and will provide whatever support is necessary. Obviously, all these factors vary depending on the manufacturer, and the organization's specific needs. The important point is to review carefully all of the advantages, benefits, potential drawbacks, costs and consequences of any system.

LIST OF SUPPLIERS

In order to help you begin the selection and evaluation process, the remainder of this chapter consists of supplier listings. We have divided current interactive video suppliers into these major areas: controllers, interface units, turnkey systems, video disc players and microcomputers. Note that the categories are somewhat arbitrary. For example, the Texas Instruments controller can be used with several video players and/or purchased independently. However, it requires the TI 99/4 microcomputer, so we listed it as a turnkey system. The same is true with other manufacturers and components. Instead of trying to list all of the unique qualifications, specifications, features and capabilities of each manufacturer, we have organized the suppliers into functional groups, since each manufacturer is continually refining its interactive technology. Each group is preceeded by a functional definition of the components in the group, and an approximate price range. (Questions about specific features and capabilities can best be answered by the manufacturers themselves.*)

---

*Readers can also refer to *Television Equipment Specification Service* (White Plains, NY: Knowledge Industry Publications, Inc., annual).

CONTROLLERS

Interactive controllers share two common traits: they can be purchased independently and operated without an external home (or micro-) computer. Controllers are easy to use and program, although some require an additional programming unit in order to segment the video material (or do sophisticated branching). The only other requirements are a video receiver and video player. Controllers range in cost from $500 to $1000 per unit.

Coloney Productions
325 N. Calhoun
Tallahassee, FL 32301
(904) 575-0691

Instructional Industries
Executive Park
Ballston Lake, NY 12019
(518) 877-7466

Informational Technologies, Inc.
6204 Benjamin Rd., Suite 209
Tampa, FL 33614
(813) 886-6193

Panasonic Co.
One Panasonic Way
Secaucus, NJ 07094
(201) 348-7000

Sony Corp. of America
9 West 57th St.
New York, NY 10019
(212) 371-5800

Videodetics
(Div. of Odetics, Inc.)
2191 S. Dupont Dr.
Anaheim, CA 92802
(714) 634-2222

TURNKEY SYSTEMS

Although the components in these systems can sometimes be purchased independently, the manufacturers in this group design systems that are compatible only with their equipment. This approach has certain advantages and disadvantages. The total system package includes a microprocessor (generally a microcomputer), an interface, video player and television receiver. The system price ranges from about $3000 to over $10,000 per package. Some manufacturers also offer lease arrangements.

Bell & Howell–A/V Products Division
7100 N. McCormick Blvd.
Chicago, IL 60645
(312) 262-1600

Emerson Electronics
8100 W. Florissant Ave.
St. Louis, MO 63136
(314) 553-3124

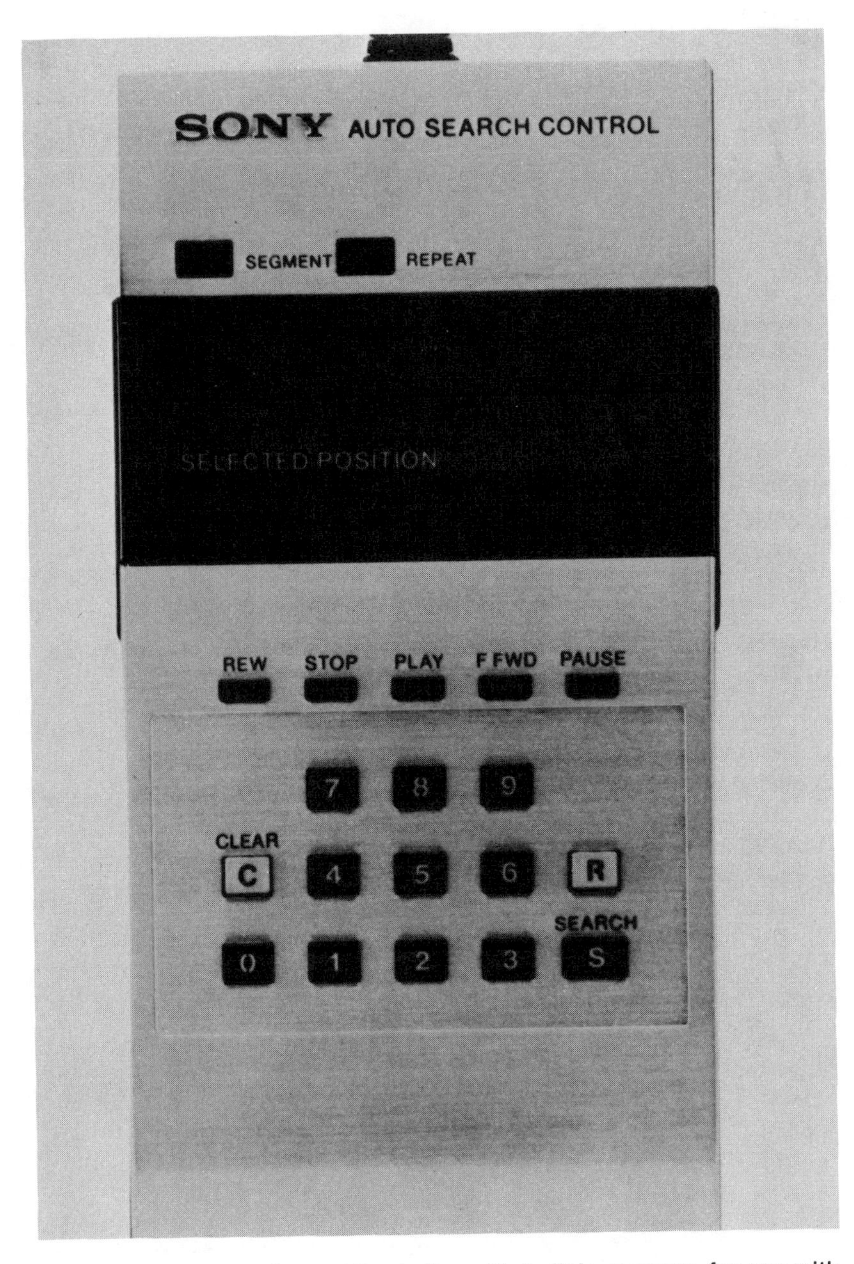

Sony Programmable Access Controller with built-in memory, for use with video cassette recorders. Courtesy Sony Video Communications.

General Physics Corp.
1 Northgate Park
Chattanooga, TN 37415
(615) 875-9688

GenTech
4101 N. St. Joseph Ave.
Evansville, IN 47712
(812) 423-4200

Media Learning Systems
1532 Rose Villa St.
Pasadena, CA 91106
(213) 449-0006

Metamedia Systems, Inc.
20010 Century Blvd.
Germantown, MD 20874
(301) 428-9160

Multi-Media Video, Inc.
3350 Scott Blvd.
Building 21
Santa Clara, CA 95051
(408) 727-1733

Panasonic Co.
(see Controllers for address)

Sony Corp. of America
(see Controllers for address)

Texas Instruments, Inc.
Box 225012
Mail Station 84
Dallas, TX 75262
(214) 995-2011

VIDEO DISC PLAYERS

There is a wide range of video disc players on the market including VHD, CED, laser optical and optical photographic systems. We will only list the major manufacturers of the laser optical video disc players, since this format is the only one which has been driven by commercially available interface systems. The current laser optical video disc format simply has the greatest potential for educational and industrial programming. Another, the optical photographic system, developed by Ardev, is unfortunately not commercially available.

Pioneer Video, Inc.
3300 Hyland Ave.
Costa Mesa, CA 92626
(714) 957-3000

Sony Video Communications
(see Sony Corp. of America, under Controllers, for address)

Thomson-CSF Broadcast, Inc.
37 Brownhouse Rd.
Stamford, CT 06902
(203) 327-7700

INTERFACE UNITS

Interface units can either be inserted into the peripheral slot in a microcomputer, or can be connected externally between the computer and the video player. No other major computer or video modifications are necessary. The interface unit relays signals between the computer and the

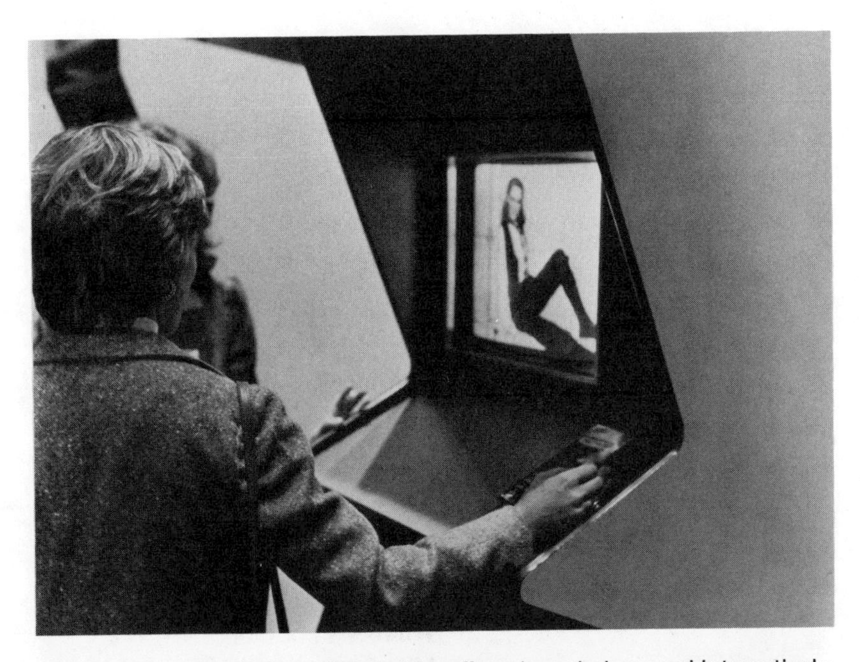

Pioneer (DiscoVision Associates) video disc player being used interactively. Courtesy Pioneer Video Systems.

The Sony laser video disc player (foreground) with built-in microprocessor for interactive application. Courtesy Sony Video Communications.

video player, enabling the computer to control the video player and make it interactive. Although interface boards can be purchased separately, manufacturers also offer authoring software to create interactive programs. The price for the interface units ranges from $300 to $500, and the authoring systems are generally an additional $300 to $500.

Allen Communications
7490 Clubhouse Rd.
Boulder, CO 80301
(303) 530-7300

Aurora Systems, Inc.
2040 E. Washington Ave.
Madison, WI 53704
(608) 249-5875

BCD Associates
1216 North Blackwelder Ave.
Oklahoma City, OK 73106
(405) 524-7403

Cavri Systems, Inc.
26 Trumbull St.
New Haven, CT 06511
(203) 562-4979

Coloney Productions
(see Controllers
for address)

GenTech
(see Turnkey Systems
for address)

Texas Instruments, Inc.
(see Turnkey Systems for
address)

Whitney Educational Services
1499 Bayshore Highway
Suite 232
Burlingame, CA 94010
(415) 570-7917

MICROCOMPUTERS

This category is possibly the most volatile of all and could easily double in the next two years. We have listed those computers that have been connected to video players in commercially available systems.

Apple Computer Inc.
10260 Bandley Dr.
Cupertino, CA 95014
(408) 996-1010

Atari Inc.
1265 Borregas Ave.
Sunnyvale, CA 94086
(408) 745-2000

Bell & Howell–A/V Products Division
(see Turnkey Systems for address)

Radio Shack
1500 One Tandy Center
Ft. Worth, TX 76102
(817) 390-3700

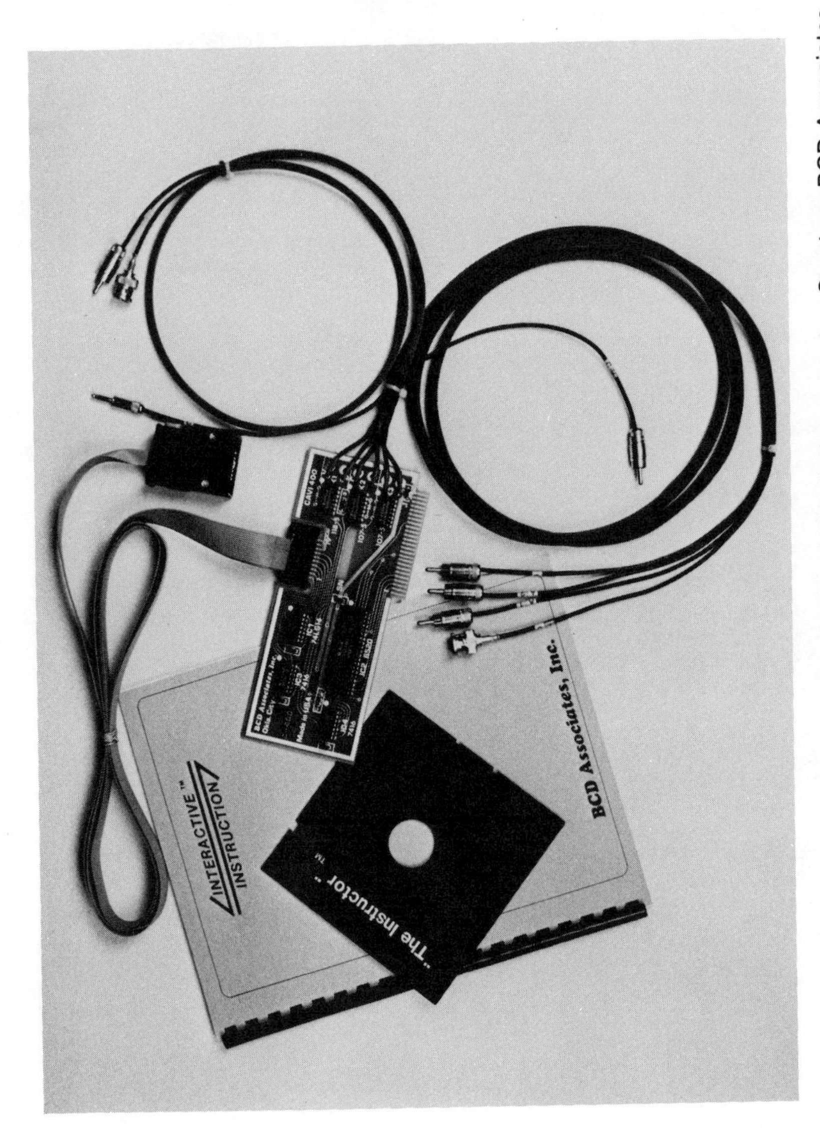

The CAVI 400 single card interface (center) fits into a peripheral slot in the computer. Courtesy BCD Associates, Inc.

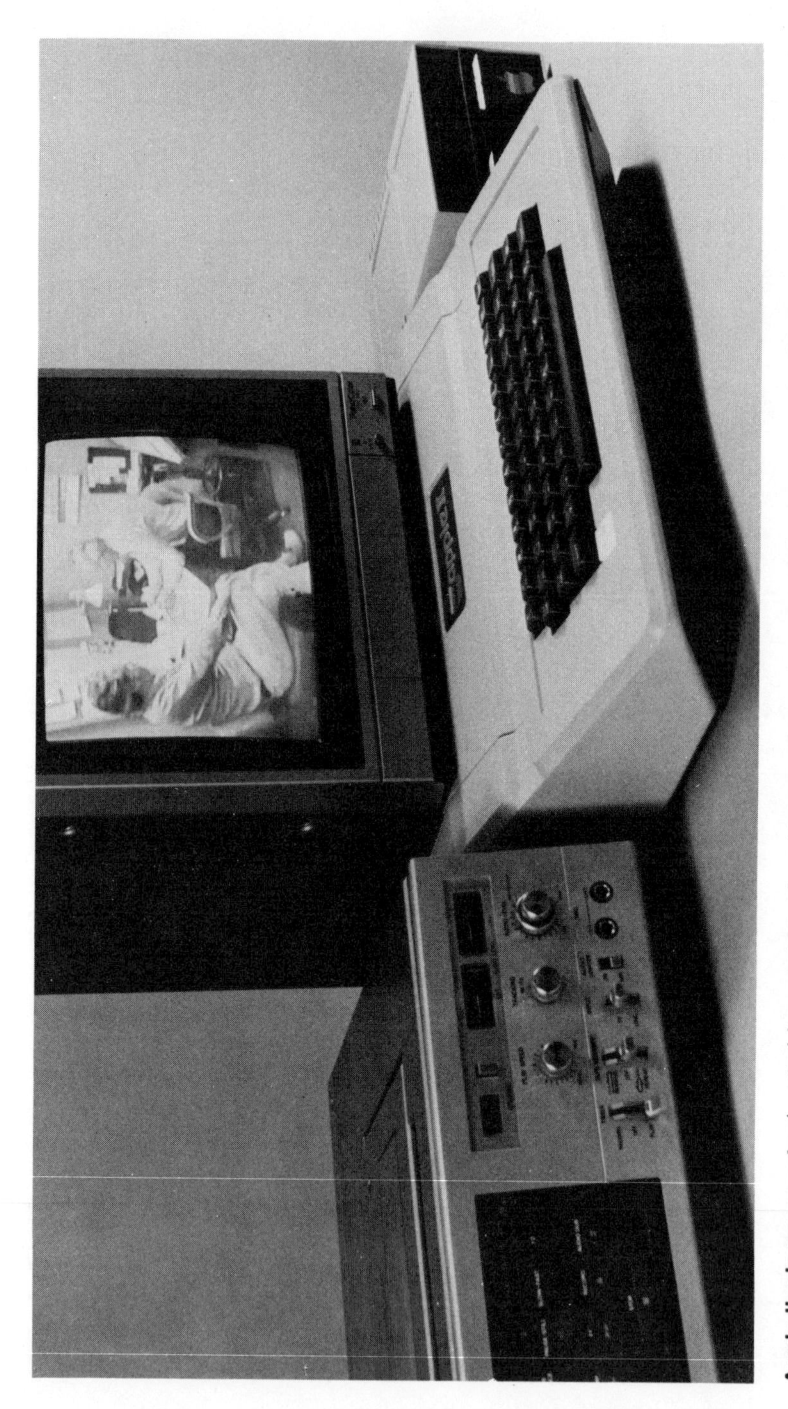

Apple II microcomputer (center) in an interactive video configuration. Also shown are the tape player (left), monitor and disc drive (far right). Courtesy Apple Computer Inc.

Solutions Inc.
3740 Colony Dr.
Suite 130
San Antonio, TX 78230
(512) 690-1017

Sony Video Communications
(see Sony Corp. of America, under Controllers, for address)

Texas Instruments, Inc.
(see Turnkey Systems for address)

WICAT, Inc.
1160 S. State St.
Orem, UT 84057
(801) 224-6400

4

Designing Interactive Video Programs

by Michael Schwarz

Your first contact with interactive video will most likely produce some anxiety within your creative psyche. Let's suppose that upper management has taken an interest in the new interactive video technology and is curious as to whether or not this new technology can be integrated into the organization. And, they want *you* to develop a demonstration program that will show how interactive video can fit into the company's overall training/communication goals.

It is at this point that anxiety begins—but fear not! As a producer of training media, you already possess most of the skills required to develop an interactive video program. And as you become familiar with the process you will see that interactive video allows you to integrate all the media skills you possess.

FIRST THINGS FIRST

Prior to developing interactive software it is necessary to have a solid understanding of the concept of interactive video as well as a thorough knowledge of the hardware. (The previous chapters of this book provide the introductory technical background for this new technology.)

The most important factor to keep in mind is that the learning path your audience follows through an interactive program is *not* linear. Rather, the student will forge his own unique path through the program (see Figure 4.1). The path is not linear but progresses along a variety of branches

Figure 4.1 Path(s) through Interactive Program Is Determined by the User's Response

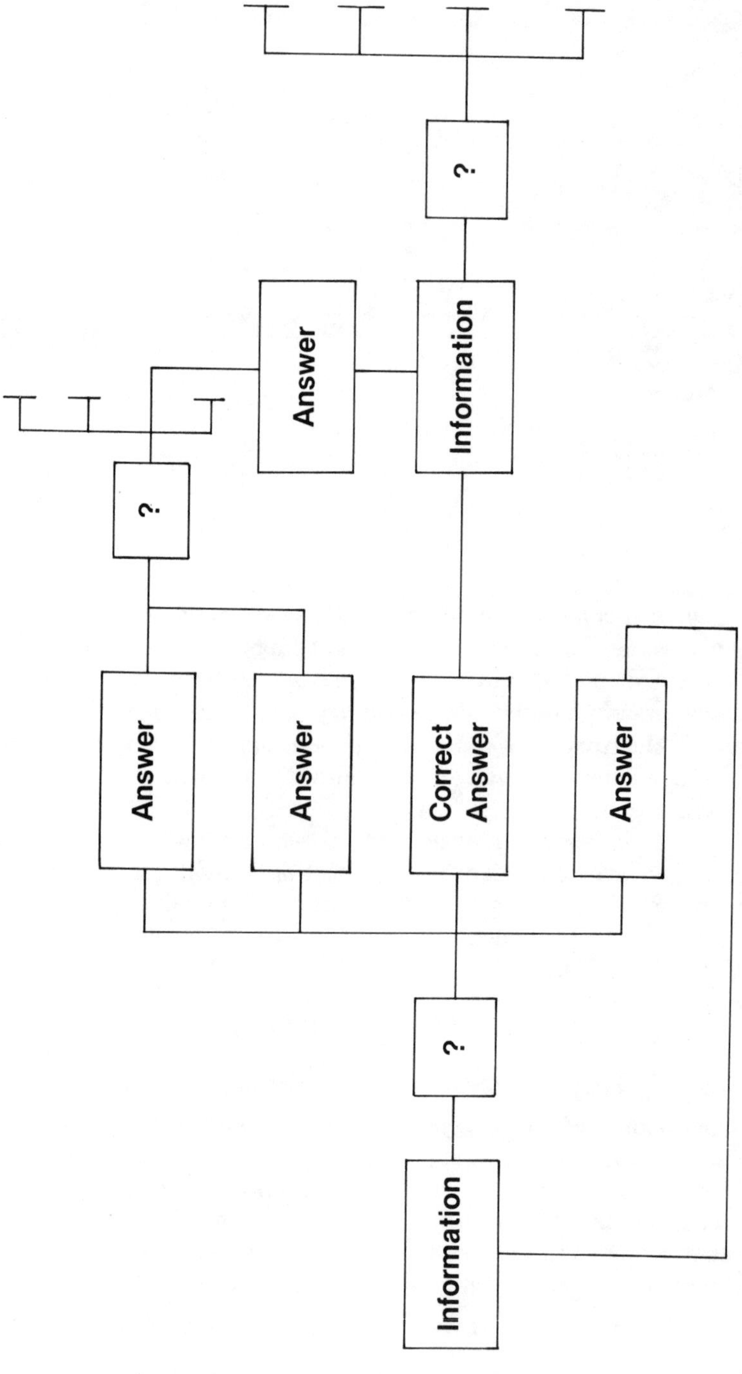

Information

?

Answer

Answer

Correct Answer

Answer

?

Answer

Information

?

depending on the student's responses. Throughout the program you will design in various points at which the student must make a decision. This interaction is what determines the learning path the student follows. Two basic design factors determine the type and method of interaction: instructional objectives and hardware capability.

Initially you must concern yourself with the interactive hardware you will be using. People who have an instructional design background may feel that emphasizing the hardware rather than the instructional content is a false priority. Granted, that danger always exists. However, it is important to remember that interactive instructional strategies have only recently begun to develop, and are a direct result of new hardware technology. Learning about the hardware should be viewed as taking inventory of new tools; your goal is to develop an instructional program that effectively meets some specific learning objectives. The hardware is simply a delivery system, a tool to support the learning objectives.

Be careful not to fall into the trap of designing a program for the system itself. Interactive video has not shifted the emphasis in instructional design; rather, it has simply removed some of the hardware barriers that have heretofore existed.

The discussion that follows is not limited to a specific system, but rather illustrates a generic system with a broad range of capabilities. There is a two-fold reason for this. First, although hardware capabilities are an integral part of the development task, they are not of prime importance. Second, the design process that follows will force you into thinking interactively. You will be designing a multi-path, branched learning video program. The various video segments along these paths will provide feedback to the student and will be accessed based on the student's own response/ interaction. This process can then be utilized on any interactive hardware system.

THE TEAM APPROACH TO MEDIA DESIGN

The development of interactive video requires a team effort: The media producer, the instructional designer (ID) and the subject matter expert (SME) all contribute their skills toward program development.

The ID is a key resource person when it comes to developing valid learning strategies. The ID is able to define the learning objectives precisely and can determine the appropriate "testing points" that will be required. The SME brings both theoretical and practical knowledge to the design team. He is a key resource person in determining appropriate remedial information.

The media producer has a two-fold role. As with normal video production, the producer is the media expert. In addition, the producer must function as the interactive video expert. Remember, interactive video is new. The producer must be the leader and guide the group through the design process. The producer's first task is to educate the design team about the nature of interactive video.

It is important to explain interactive video from the student's point of view, and to tread lightly when it comes to explaining the hardware. Too much technology can be intimidating and the design team might also begin to encounter the early stages of panic. But be sure to explain the system's capabilities, using concrete examples of how interactive video will help meet specific learning objectives. It is particularly important to explain the concept of random access, or the design team may slip back into traditional linear thought patterns.

You can begin by presenting a video program with which everyone is familiar. Then, by stopping the program at various points, you can explain how an interactive video program on this same subject might have looked.

At this initial meeting be prepared for an information overload. The team members will probably start discussing "teaching machines," "programmed texts," "computer assisted instruction (CAI)," "computer managed instruction (CMI)," and so on. And that's good, because elements of all these methods are used in interactive video.

The main objective of this first meeting is to get the design team to think interactively, and not in the traditional thought patterns associated with passive, linear video programs.

DEVELOPING A MEDIA STRATEGY

Every company has its own unique method of selecting topics for media development. Some follow a detailed instructional system design (ISD) model, while others will produce a program for whomever has the budget. Most organizations probably fall somewhere within that broad spectrum.

Topics that deal with sequentially ordered tasks are especially suited to interactive video. These can include:

Maintenance training. Where a student must follow a lock step procedure (i.e., step one must be accomplished before proceeding with step two, etc.). Throughout the training procedure an interactive program can test the student's mastery of the previous steps.

Troubleshooting technique. Here, the student must learn a logical approach to identifying a problem and correcting it. Interactive video can lead the student through the "discovery" process associated with troubleshooting techniques. At appropriate points the student can branch ahead and try a solution. If his solution is correct the student will continue on through the program and resolve the malfunction. If incorrect, the student will be branched back to the original troubleshooting sequence.

Other topics that can be developed for interactive video programs, but which require more design effort, include those that involve judgment skills, such as management skills training. For example, let's say that a manager must learn to adapt his approach to a particular subordinate. There are no right and wrong approaches; rather there are multiple outcomes based upon the manager's particular response. This type of topic development, in which there are multiple alternative endings and results, is a more difficult type of interactive program to develop. Basically, this is because there is no readily apparent internal structure.

GETTING STARTED: THE MEDIA STRATEGY STATEMENT

Let's assume that a topic for an interactive video program has been identified. To begin, you should convene a meeting of the design team. The first task that must be accomplished is development of a *media strategy statement.*

A media strategy statement is, essentially, a specification sheet for the interactive program you will develop. It will set the limits for your program, and as such must contain very specific information.

Begin by examining all existing instructional materials that deal with your assigned topic. The ID and SME should be able to explain all the objectives that are in the programs. After such a review you will be able to narrow down the objectives that can be achieved by an interactive program.

Setting Objectives

The ID will be able to assist you in writing sound objectives for the program. The objective should be student-oriented and describe what task the student can perform given a specific set of conditions. In addition, the objective should include criteria for evaluating accomplishment of the task.

A word of caution: interactive video can do a lot of things, but it is not a panacea. You will be doing your company a disservice by developing an interactive program that attempts to accomplish too many objectives, since this type of shotgun approach almost always leads to failure.

Using Existing Materials

Once you have determined the objectives of the program, you can begin looking for existing media materials that deal with your assigned topic. These can take the form of slides, file film footage, programmed texts that deal with your topic, etc. It may be possible to produce an interactive video program as an adjunct to a programmed text. For example, a page of a programmed text on flight training might contain a question like this:

> You are flying in actual instrument conditions and become momentarily disoriented. Your aircraft has assumed an unusual attitude. The proper sequence for recovery is:
>
> A. Lower the nose, level the wings, adjust airspeed . . .
>
> B. Level the wings, lower the nose, adjust airspeed . . .
>
> If your answer is A—turn to Page 32
>
> If your answer is B—turn to Page 36

An interactive video program based on this text would elicit a response from the student at this point, rather than have him simply turn a page. In addition, the student would actually see the results of his particular choice. The incorrect response would be a video segment showing, from the pilot's point of view, the sequence of recovery chosen by the student: e.g., the aircraft has entered a stall and possibly goes into a spin! This visual response is much more effective than mere words on a page. The narration would further amplify the correctness/incorrectness of the student's choice. The student could then view some remedial video or be branched back to the programmed text.

This example illustrates what can be accomplished with existing materials. Interactive video's random access capability provides new tools and allows you to integrate media in new and exciting ways. There are producers who have been reluctant to mix media (film, video, slides, books) in traditional linear video programs, citing obvious visual quality differences. These differences can include static development of slides juxtaposed with dynamic motion of video or graininess of film intercut with nice crisp video. But random access overcomes this problem. Scene A (video) followed by scene B (film) followed by scene C (slides) is no longer a restriction, since the material will not be presented in a linear pattern. As a matter of fact, the learning event referred to in scene B might not even be a video segment; quite possibly, it could be a workbook exercise.

In performing your analysis of existing materials remember that the various media will not be integrated into a straight line program. Certain elements of your program's objectives might require only still pictures instead of motion sequences. Interactive video will allow cost-effective, instructionally sound media integration.

Modifying Existing Programs

Certain considerations, such as a limited budget, necessitate modification of an existing video program to make it interactive. This alternative is also appropriate when a very basic form of interaction is required. For this approach, simply develop a series of multiple-choice questions throughout the program. Then use remedial video segments to guide the student back to that part of the program covering the appropriate material. But you should be aware of some inherent problems with this approach. The development of teaching points and the instructional design efforts that go into a linear program differ from what is required for an interactive program. All the "retrofitting" in the world will not yield a program as effective as one designed from scratch.

Any retrofitting requires you to consider some basic questions: Is the same talent available? Will you have any problems matching the remote filming locations, etc.? How are you going to treat the topic? Is your subject matter being presented for the first time or is your interactive program a follow-up to a classroom lecture? What is the audience's prior level of knowledge and experience? This last step is important because your interactive program must be designed to accommodate a multilevel audience. Progress through the program will be based on each student's abilities and responses.

Develop a Strategy Statement

The next step is to develop a strategy statement. Figure 4.2 is a media strategy statement for modifying an existing linear video program. The strategy statement for an original interactive video program, discussed earlier, would include the same areas of discussion (objective of program, analysis of learning objectives, etc.). The strategy statement should describe the type and degree of interaction you plan to use. Will all of the students begin at the same point and then branch as they proceed through the program? Will you present a bank of questions at the beginning to determine where in the program your student will begin? Will multiple-choice questions be used? Will the various interactions eventually lead the students to the same ending segment? Or will you have alternate endings? Will you incorporate drill and exercise activities into the program? Does your hardware system allow for conditional and unconditional branching? (This last question uncovers a whole range of variables, which are considered in more detail below.)

These questions will shape the structure and look of the interactive program. The whole team should have input into the answers. The ID can recommend the type of learning activity best suited to each objective; the SME will have practical experience that will suggest other types of interaction; and the media expert can determine how the various hardware capabilities can be utilized to implement the desired type and degree of interaction.

Figure 4.2 Strategy Statement for Modifying Existing Program

Objective of Program

After successfully completing this programmed unit (video tape and workbook) the student will be able to identify an author's *style* and *tone* after reading a selected passage of prose.

Analysis of Learning Objectives

The design of this program covers three learning steps:

1. The student must learn new vocabulary.
2. The student must learn new definitions and identify key elements of those definitions from the examples provided.
3. The student must transfer the knowledge gained in steps 1 and 2 by applying it to new prose passages and determining the style and tone.

Steps 1 and 2 are the easiest to present and learn. They are simply an identification of facts (definitions). Step 3 is a larger "jump" for the student, because he must use his previous knowledge and make a judgment to determine the answers.

As this answer is a "best" answer, there is a subjective "gray area" the student must conquer.

Treatment of Topic

The tape presents the new vocabulary and definitions via exemplary vignettes. All the "styles" are presented and then followed by a mini-review. "Tones" are presented in a similar manner. The student is then asked to identify the style and tone of two prose passages.

Evaluation of Existing Program

It is an extremely well produced tape geared to an adult American audience. The resulting instructional pace is too rapid for our students. Moreover, our students will not have enough opportunities to apply their knowledge to prose samples if this tape is used, as is.

Strategy for Implementing Program

The required information can be presented to our students and absorbed by them if that information is presented in smaller steps. A student should also be allowed/required to apply his knowledge to prose passages *throughout* the program.

The existing tape segments will be used to present the new vocabulary, along with the definitions and examples. Our insert will reinforce the definitions and then require the student to apply this information to a sample, thus inducing a transfer of knowledge.

This step-by-step learning and confirming process will better prepare the students for the final test samples as presented in the existing tape.

Conditional and Unconditional Branching

Unconditional and conditional branching are key distinguishing features between interactive video and linear video programs—and there is a major difference between these two types of branching. In an unconditional branch, there is one and only one way to proceed. For example, if a student responds to Question 10 by selecting A, the system will always branch to a specific video segment, Segment #28. In this type of system all students who respond to Question 10 with A will go to Segment #28.

In a conditional branch this is not true. For example, let's say a student has responded to Question 10 with response A. At this point the system begins to evaluate a set of conditions that was previously determined by the program designer. This could include: the number of previous correct/incorrect responses; the particular response to the previous question; the time it took to make the response, etc. Based on the evaluation of these conditions the system will then branch the student to one of several segments. This could be the same Segment #28, as above, or it might be another segment, such as Segment #29, or Segment #30. The branch depends on what conditions were met.

A computer programmer might explain these two types of branches as follows:

Question 10:

Unconditional: If A, then go to video Segment #28.
Conditional: If A, then if (set of conditions) go to
Segment #28
else if (another set of conditions) go to
Segment #29
else (a third set of conditions) go to
Segment #30.

Unconditional and conditional branching techniques are the basis for the three levels of interaction described in Chapter 1: basic programmed learning, programmed simulation and exploratory simulation. Obviously, the ability to incorporate conditional branching implies that the system has a pretty sophisticated degree of intelligence and memory. Typically, this means that the hardware system includes a minicomputer.

DEVELOP TOPIC OUTLINE

Mainline Sequence

At this point you can begin implementing your strategy statement.

Based on research and consultation with the subject matter expert, outline the subject matter. First develop the *mainline sequence* of the presentation; that is, proceed as if you were developing a traditional linear video program, and determine the presentation sequence you will follow. Now examine this sequence and determine where there are natural breaks in the material.

For example, if you are presenting a maintenance activity, the many subtasks required to perform that activity will form logical breaks. By doing this you can develop the instructional segments required to achieve your program's objectives. Further, you can examine these segments and determine whether or not they will be video, computer interaction, text, etc..

Instructional Segments

Next, carefully analyze the instructional segments that make up the mainline sequence. Determine which segments you feel will be difficult or confusing for the student and begin outlining remedial segments that will be used to amplify or reinforce the mainline sequence. In addition, determine the type of remedial information that would clarify the problem areas. When outlining these various segments be sure to specify how much detail you want to include. You must be sure that you include all the teaching points necessary to support your instructional objectives.

Degree of Interaction

Now begin to design the type and degree of interaction that you specified in the strategy statement. Where will the decision/testing points be? How are you going to determine whether the student will proceed with the next instructional segment or will see some remedial information? Will you use a single multiple-choice question or a series of questions? What criteria will you use at each testing point to determine the path that will be followed? In developing appropriate questions, rely heavily upon the SME and ID. They have the experience and knowledge required to write effective questions.

At this point you have a fairly detailed outline covering the subject matter (whether instructional, remedial, etc.) you intend to develop. However, it is still just a list of elements. You are now ready to begin formalizing the program. And, like it or not, that formalization takes the form of flow charts. Until now, the process has closely paralleled what you might do for a traditional linear video program. Now you will begin encountering tasks that are unique to interactive video program design.

STARTING THE DESIGN PROCESS

In traditional linear video programming the student has no choice as to what path he may take through the program, since there is only one path. Every student starts at Point A, sees the same exact sequence of video, and arrives at Point B.

As you know, interactive video is somewhat different. Students may start at Point A, or Point X. They may see the same segments of video or different segments, and may arrive at Point B or Point C, depending on their responses. Therefore, the next task is to design the various paths that the student may follow, by developing flow charts that show all possible sequences.

During the rest of the design process you will create several flow charts. It is important to remember that each flow chart is created for a specific purpose. If you attempt to consolidate several flow charts to form one multipurpose flow chart, you will almost always create an illegible document—and this will generate mass confusion for everyone concerned.

Creating the Instructional Flow Chart

The first, and perhaps most critical, flow chart to develop is the instructional flow chart. This diagram shows the sequence(s) of *instructional events* that will occur throughout the interactive video program. This document is based on the information you developed in the topic outline. It documents all of the paths available to the student *and* all of the instructional activities in which he or she might engage. As such, this chart will include non-video as well as video information. It might include workbook exercises, computer interaction, performance activities, and anything else that is necessary for the interactive video program.

The instructional flow chart shows the mainline sequence of instruction as well as the various remedial loops. This flow chart must be as detailed as possible because it serves as the master blueprint for the entire interactive video program. This document is especially helpful since it can be used to review the overall program.

Thus far, all you have done is design work. Not a word of script has been written and, certainly, nothing has been committed to video. Prior to moving into the development phase (and the subsequent expenditure of resources) you should take this document and obtain whatever preliminary approvals are normally required by your organization. You can easily "walk" someone through the program by using the flow chart and outline.

Now let's walk through the instructional flow chart shown in Figure 4.3.* Non-video instruction is depicted above the line, video instruction, below.

(1) Each student sees this introductory video segment. This segment explains the overall objective of the program and ends by directing the student to the technical order (T.O.) where he is to read the first part of the maintenance procedure—grounding the aircraft.

(2) After reading the T.O. the student returns to the video and is shown that part of the procedure he has just read (Procedure #1, in Figure 4.3). The student sees how the maintenance specialist follows the tech order in performing his tasks. This video segment ends by directing the student to review the T.O. and then answer some questions about the grounding procedure from the workbook (3). Upon completion of these questions, the student presses a button on the keyboard which produces a "call instructor" message. The instructor checks the student's answers and allows the student to see a multiple-choice question on video (4).† The student answers the question by pressing a button on the responder.

(8) A correct answer causes the student to see a video segment that provides reinforcement. The segment then ends by directing the student to the tech order to read the next portion of the procedure.

(5) An incorrect answer branches the student to a video segment that visually provides remediation. The segment then directs the student back to the T.O. and workbook (6) for some more fill-in-the-blank questions. After an "instructor check" the student then is given a second chance at the video segment's multiple-choice question (7).

On this second pass, an incorrect answer would elicit a "call instructor" display. At this point the course administrator would personally provide the remediation. The correct answer branches the student to the video seg-

*Figure 4.3, an instructional flow chart, is based on an interactive video program produced by Northrop Worldwide Aircraft Services, Inc., in which the author functioned as media producer. As such he was responsible for overall program design, in addition to acting as writer/director on this project. In this program the student is guided through a typical aircraft maintenance procedure—single-point refueling of a T-38 aircraft. An objective of this program was to help the student visualize the tasks dictated by the written word. In addition, the student was forced to extract specific information from the written procedure. For this program we used a video tape, workbook and the appropriate USAF Technical Order (T.O.) which is a maintenance manual.

†Note: If the Northrop hardware system had some intelligence and character generator capability (such as an Apple computer would have) we could have eliminated the workbook. All workbook exercises would then have been simple computer aided instruction (CAI) exercises on the video monitor, which would be checked by the computer and automatically followed by the video segment with the appropriate multiple-choice question.

Figure 4.3 Instructional Flow Chart

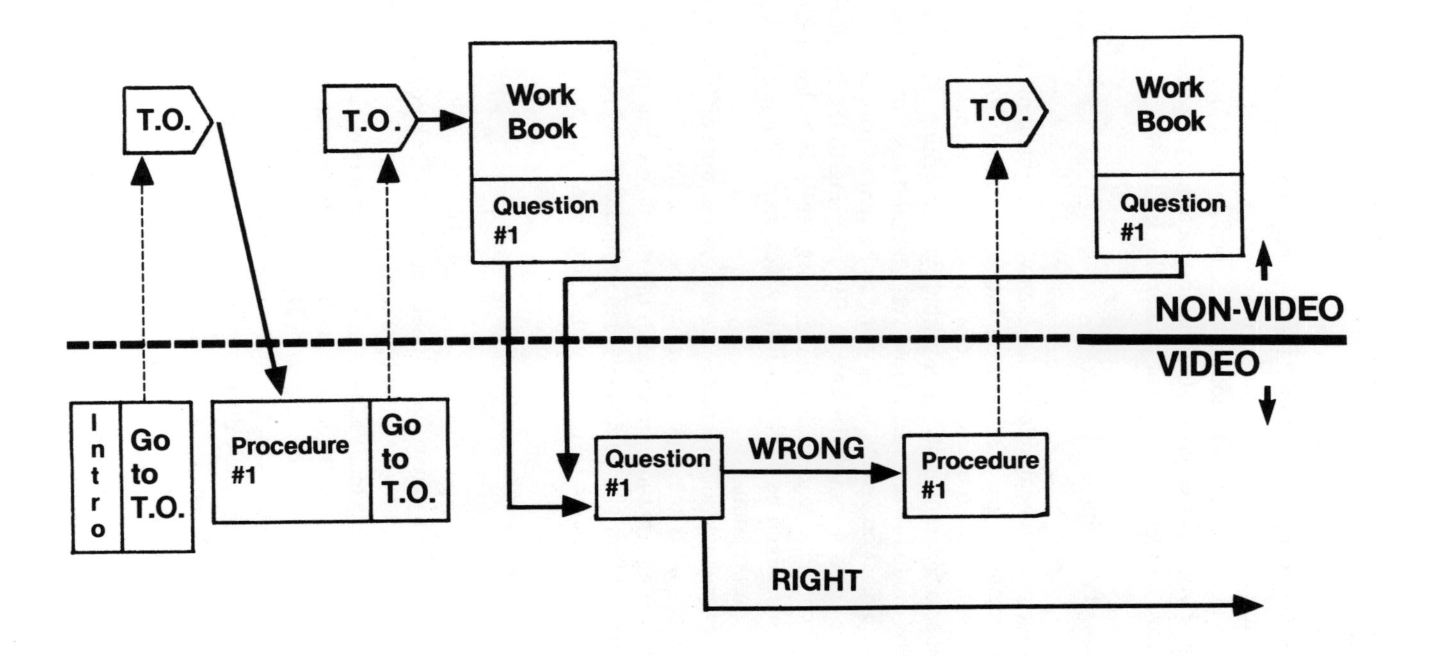

ment that he would have seen had he answered the question correctly originally.

The rest of this instructional flow chart, and the program itself, continues in the same manner until the refueling procedure is complete. Not all students complete the program. Those who get bogged down (usually with the vocabulary) are given additional remediation on another occasion.*

Once you receive approval on the instructional flow chart you are ready to move into the development phase; this begins with scriptwriting.

Writing the Script

A script must be developed for each video segment of the program, and these segments are identified on the instructional flow chart. Initially, there is no need to be concerned with specific sequencing. Treat each segment as a separate entity. (Remember, you're writing video segments that will be randomly accessed.) However, you must take into account whether the segment you are writing is going to be used as part of the mainline sequence or as a remedial segment. (By referring to the instructional flow chart you can readily determine a segment's intended use.) Mainline segments should be written as if they were parts of a regular instructional television program.

The remedial video segments are somewhat more difficult to write since a student views this type of segment because he did not correctly understand the information presented in the mainline segment. Refer to the instructional flow chart and analyze the content of the appropriate mainline segment. There are various remedial strategies that you can employ.

Remedial Strategies

The simplest type of remediation, and one that will not require additional video, is to show the mainline segment to the student a second time. Instructionally, however, this approach usually is not the most efficient or effective. More often, you will have to create a new video segment whose sole purpose is to provide remediation on a particular teaching point that was within the mainline segment. Try to write this type of segment using smaller learning steps that include more detail than in the original mainline sequence. A completely different set of video shots, or editing the original shots in a different sequence, may also serve to *visually* clarify the par-

*Note: If you're just getting into interactive video try to begin by using a simple type of interaction as described above. It's always easier to walk before you run.

ticular training point. If you follow this course, your ID and SME can be of assistance and can help you write this second remedial explanation.

Writing video segments that remediate wrong answers is a tricky process —writing logical wrong answer segments is not as easy as you might think. Research tools, such as maintenance manuals and technical dictionaries provide correct answers—e.g., the proper way to change a spark plug, or the proper definition of a word. They do not indicate how *not* to change a spark plug, or how to define a word improperly. Therefore, it is difficult to assess why a student might have selected a wrong answer or option. For wrong answer remedial segments rely heavily on your SME—his experience will help you write effectively. The SME can also suggest the rationale for the incorrect option.

Be judicious in the use of remedial segments for multiple-choice questions. You don't need to write a unique video segment for every wrong response. For example, in the Northrop program discussed earlier, there were three choices in each video multiple-choice question. The two incorrect responses each used the same remedial segment, because we felt that the content was so basic and generic that one common remedial segment was sufficient. Your decision in this matter should be based on the content of the question and the particular mainline (or instructional) segment. Be selective in writing and specifying the various remedial segments.

Frequently, video segments will refer to non-video items such as workbooks, computer drill and exercise, etc. At this scriptwriting stage you should at least develop an outline of these non-video items—and be detailed. Media integration is much more effective if your video segment is specific—i.e., if it says "Turn to Page 7 in your workbook" rather than "Refer to your workbook." Your script can contain this type of detailed referral information, *only* when the non-video items have been outlined in detail.

Writing the Questions

The ability to write a valid test question is an art unto itself. If you use a video segment to set up a question, you will have to be creative and careful. One of the problems that test writers encounter is that an improperly written question can easily give away the answer. Tread carefully when using video for this same task. Remember the camera sees all, and your visual set-up for the question can easily give away the answer. So be cautious and don't provide too much information.

You should now begin to get the required approvals on your draft script. In the meantime, you are going to develop another flow chart!

Develop Interactive Video Flow Chart

The purpose of the interactive video flow chart is to determine the format of your video source material *only*—the format of non-video materials is not included. This flow chart will show:

- The sequence of video segments as they are to be recorded on the final video tape.
- The content of each video segment.
- The various places the video will be stopped. These stops are decision points (see Chapter 1).
- The direction the video is to be shuttled after each decision point. That is, after a particular video segment has played and is stopped, what segment is going to come next? And, as you know, there are a number of paths possible at each decision point.

In determining the sequence of video segments there are both physical and instructional factors to consider.

Access Time

Presently, there are no hardware systems that provide instantaneous access time. A video disc player can take as long as 5 seconds, while a VHS or Beta machine can take a couple of minutes, front to back. Your task, then, is to minimize the access time. The shorter the shuttle time, the quicker the feedback to the student. Segments need not be sequenced in order. For example, a remedial video segment physically can be placed before the respective mainline segment in order to reduce search time.

Get together with your instructional designer and determine which answer segments should have shorter access time—those for right answers or wrong answers. In the Northrop program, we felt that it was acceptable for the student to wait somewhat longer for the correct answer reinforcement. We sequenced the tape so that the shortest shuttle time was for the remedial wrong answer segment. (For your first few programs, don't get overly concerned with this sequencing step. This step takes on importance when your program is complex and there are large numbers of video segments and branches. By then you will have had more experience, and planning the sequence will become second nature.)

Probability of Response

When it comes to segments that are answers to multiple-choice questions, those answers with a greater probability of being chosen should be placed closer to the question segment than the other answer segments. This, again, will reduce search/shuttle time.

To create the interactive flow chart simply refer back to the instructional flow chart. Begin placing all the video segments in your chosen sequence. After each segment the video will be stopped for a decision point. By reviewing the content you can readily determine where the video may be shuttled to. Figure 4.4 illustrates the interactive video flow chart developed for the Northrop program discussed earlier. It diagrams all the video segments referred to in Figure 4.3 (i.e., the segments appearing below the line).

Segment #1 is an introductory segment, and has only one path, which leads to Segment #2. In Segment #2, the student views a procedure (here, grounding the aircraft—this segment corresponds to (2) in Figure 4.3). The next video segment is #3, a multiple choice question. This decision point has three responses—two incorrect, and one correct—which are shuttled to two different video segments. The two incorrect responses, A and B, are branched to Segment #4, a remedial segment. The correct answer branches the student to Segment #5, which provides reinforcement and leads to the next sequence in the program. Remember that this flow chart is concerned with video source materials *only.*

Assemble Final Script

With the flow chart complete, you're one step closer to actual production. At this point you should have an approved draft script. The script itself is not necessarily in any particular sequence. You simply write a script for each video segment, and create a final script by conforming the draft to the interactive video flow chart. Arrange your script so that it now reflects the same sequence that appears on the interactive video flow chart. The two documents *must* agree in sequence. Figure 4.5 illustrates the final script for the Northrop program. If you refer to Figure 4.5 you can see that this final script is like any script you would normally develop (except for the instructions in line #13, which were necessitated by the Northrop hardware system).

In assembling the final script you may be required to make some additions in both the audio and video columns. The final script needs to indicate any special accommodations required by your particular hardware. As you can see, the Northrop system needed 10 seconds of black before

Figure 4.4 Interactive Video Flow Chart

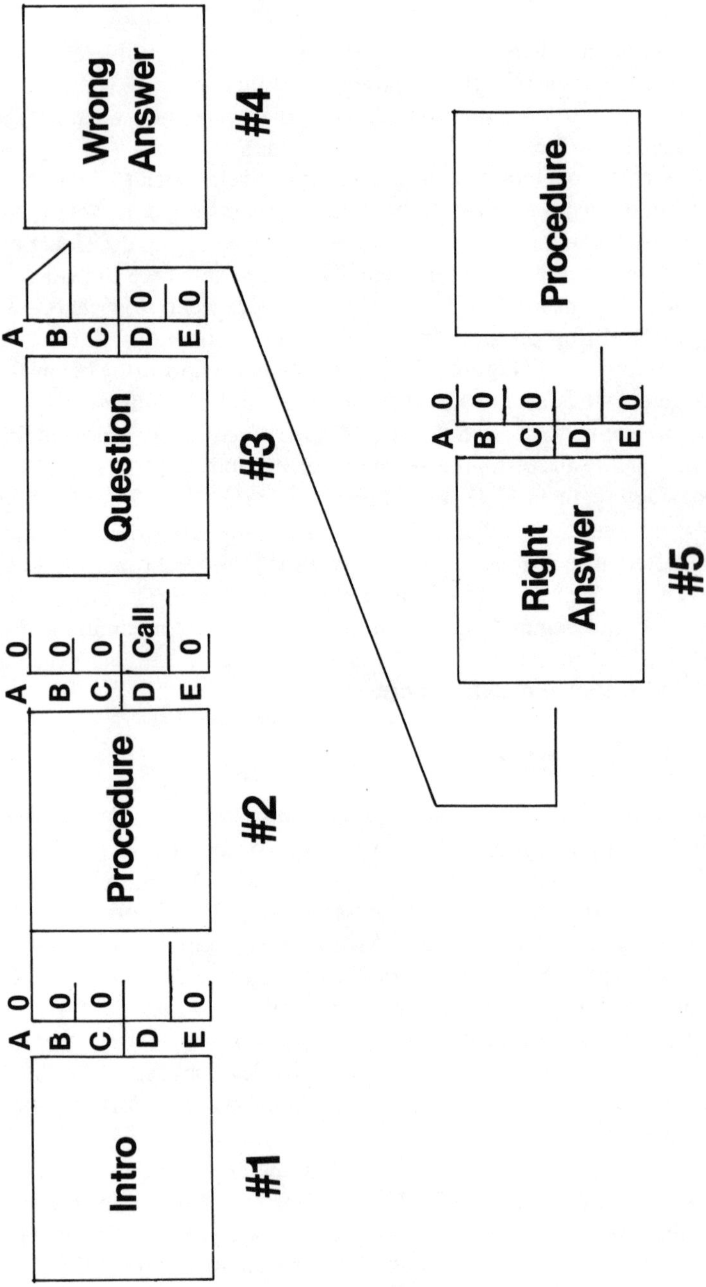

Intro
#1
A 0
B 0
C 0
D
E 0

Procedure
#2
A 0
B 0
C 0
D Call
E 0

Question
#3
A
B
C
D 0
E 0

Wrong Answer
#4

Right Answer
#5
A 0
B 0
C 0
D
E 0

Procedure

each video segment. Other systems may require a blank audio channel 2 at the beginning of the program. All systems require something. Be sure that this is noted in the final script. With final script in hand, you can now make your way to the TV studio!

PRODUCTION

As always, production simply follows the final script. The production crew does not have to be concerned with the branching. The script, because it is segmented, may not make immediate sense to the production team, but a brief explanation should answer any questions. A detailed explanation of production is covered in Chapter 6. The only point I would like to call attention to is control track referenced hardware systems. These systems reference the tape by counting control track pulses—therefore, some of these systems require electronically exact dubs. If you have such a system it would be a good idea to place a "start here" key at the head of your edited master tape. This can serve as the common initial reference for all copies of your program.

SYSTEM PROGRAMMING

This final phase of the development process is probably the one you will find least appealing. In it you must develop the computer-type programming required to make your hardware system "smart." The complexity of this step depends on what system you use. Programming computer interaction activities will not be discussed in detail here since each system has unique requirements. However, authoring, or programming, must occur before your entire interactive video program is complete.

The type of programming involved in this step is really decision logic. You will be telling your hardware system where the various video segments are located on the tape or disc and when the segments are to be accessed. Essentially, you will be "mapping" the video source material by inputting geographic coordinates and instructing the system which road to take at any intersection. Programming languages for various hardware systems differ. There are, however, some common requirements for all systems.

Remember, a linear video program always starts at the beginning of the tape and plays until the end of the program. The random access of interactive video requires many different start and stop points. Therefore, you must first indicate the start and stop of all the video segments. Do this by writing down reference numbers where the segments are going to be located. In our system, the ABC 2000, which was control track referenced, we put the tape into the player and played it straight through. Then we

Figure 4.5 Sample Portion of Final Script Form

Final Script Form | **Page 12 of 28**

ABC 2000
Title: Maintenance Procedures: Single-point Refueling

Approved by: ______
Date: ______

VIDEO		AUDIO
CU of appropriate T.O. page.		WELL DONE! ACCORDING TO SECTION
	1	3-4, STEPS A-C OF THE T.O. THE FIRST
Begin shortened grounding	2	GROUNDING STEP WAS TO ATTACH THE
cable operation.	3	GROUND CABLE FROM THE AIRCRAFT
	4	TO THE STATIC GROUND. YOU CHOSE
	5	THE RIGHT ANSWER BY READING THE
Last cable connection.	6	T.O. AND OBTAINING THE CORRECT
	7	INFORMATION. LET'S CONTINUE THE
	8	SINGLE-POINT REFUELING PROCEDURE.
	9	WHEN THE TAPE STOPS, GO TO THE T.O.
KEY " Read ____ T.O. D	10	AND READ STEPS D THROUGH I. THEN
through I".	11	PRESS BUTTON ______ TO BEGIN.
HOLD KEY	12	(10 SEC. NATURAL SOUND)
X-FADE (10 sec. black)	13	
	14	
WS of scene.	15	VOICE OVER

Figure 4.5 (cont.)

Technician begins to check	16	(5 SEC. NATURAL SOUND)
these valves, drains.	17	AFTER THE GROUNDING STEPS HAVE
	18	BEEN COMPLETED, THE MAINTENANCE
CU of each and KEY as	19	SPECIALIST CHECKS THE . . .
mentioned.	20	
"Fuel tank drain."	21	FUEL TANK DRAINS . . .
	22	

Source: Northrop Worldwide Aircraft Services, Inc.

read the beginning and end points of each segment from a counter on the hardware. At times you may want to access specific scenes within a video segment. If so, you will also have to record the beginning and end of the appropriate scenes.

Develop a Branching Chart

Time for one more chart. You need to develop a branching sheet. Refer back to the decision points that you incorporated into the interactive video flow chart. This new branching chart is simply an expansion of these decision points. Figure 4.6 depicts the branching chart developed for the Northrop program. It shows all the decision points incorporated in Figure 4.3 and the number of the video segment to which the tape will be branched. The actual path that will be chosen depends upon the student's response. This branching sheet merely shows all the paths that *must* be programmed into the hardware system.

Now that you have the segment locations and the command logic (shuttle, or branching, paths) on paper, you have to physically program the system. To do this you will have to be familiar with the programming manual for your system and convert the above information into the language your system understands. This task is not as mysterious and intimidating as you might believe—it's just different.

Figure 4.6 Branching Chart for an Interactive Program

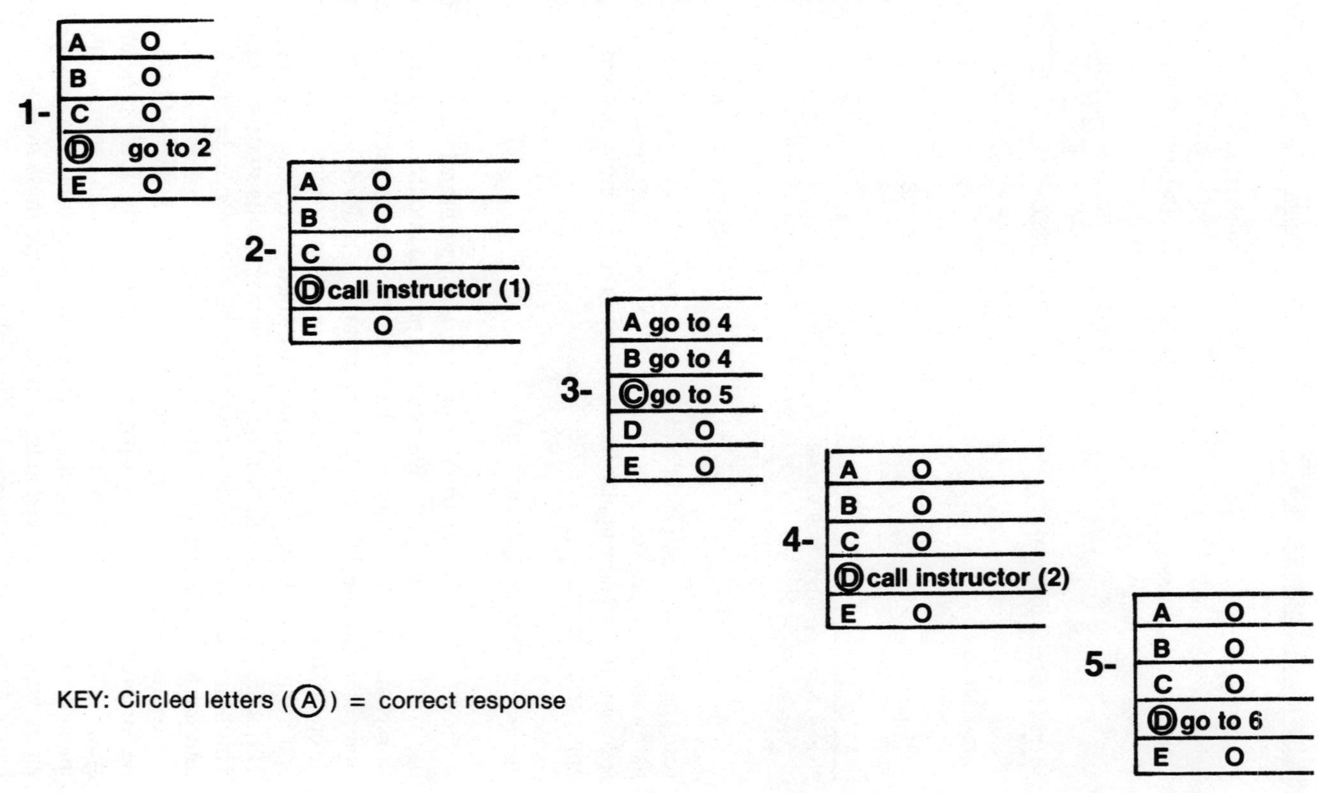

KEY: Circled letters (Ⓐ) = correct response

The last part of this task is to verify the programming. This means that you must choose *all* the possible branches at *all* the decision points. This is the only way you can be certain the video will shuttle to the proper segments. This is a tedious task, best accomplished with large amounts of coffee.

ASSOCIATED MATERIALS

The final step is to develop all the associated non-video items. This means that all the CAI exercises, workbooks, etc., must be completed. In addition, you must determine if any additional materials are going to be required to administer the program. If so, this is the time to complete them.

SUMMARY

At first, the number of tasks required to develop an interactive video program may appear to be overwhelming. But as you gain experience you will see this is not true.

The design process itself is not linear. Once you have developed an instructional flow chart you can actually network the task. Script development, developing the associated instructor/student materials, the interactive video flow chart, and the branching sheet can all be accomplished at the same time to reduce overall development time.

Initially, I would recommend that you follow the sequence as outlined in this chapter. This may lengthen the development time for your first few interactive video programs. However, there are too many variables to consider to assign a specific time frame to each step. It is like trying to specify how long it should take a user to complete an interactive video program! The amount of time it will take to develop a program depends on how much experience the media producer has with interactive video.

Interactive video offers a creative challenge never before available. To paraphrase Robert Frost: You will be designing video programs that address the roads not taken—as well as the many detours along the way.

5

Instructional Strategy and Evaluation

by Kenneth G. O'Bryan

THE CASE FOR EFFECTIVE INSTRUCTIONAL DESIGN

Instructional design will make or break interactive video as a commercial proposition. The new video tape and full-function video disc technologies must capture the institutional market or they cannot be expected to sell. Without an instructional design component that organizes and structures the content to exploit their interactive capacities fully, these systems offer few advantages over standard video tape and non-interactive video discs.

Despite the brilliance of these technologies, coupled with the almost universal fascination television holds, we must learn how to use interactive video systems to teach. Furthermore, these new systems must prove more effective, both as teaching and learning tools, than current technologies. Otherwise, they will flounder in a marketplace already overburdened with expensive, ineffective toys and undersupplied with discretionary dollars.

Not only must interactive systems discover the elusive art of teaching without human mediation, they must find it with a medium that has legions of critics, some supporters and few effective practitioners. Indeed, instructional television, for all the time and millions of dollars invested in it since the early fifties, has not captured its potentially captive audiences. Most critics and some researchers identify instructional television's primary problem as the creation of passivity in audiences. Traditionally the design of linear, timebound educational programming has been constrained by broadcast schedules, school timetables, administrative structures, teacher resistance, and sadly, the target audience's apathy for television programs lacking the mass appeal and production budgets of commercial counterparts. Despite some excellent production and fine instructional design in children's programming, very few series are widely used in classrooms. In adult education virtually no major programs are used regularly.

Even under the more favorable conditions of nonbroadcast industrial applications, in which special programs are provided for carefully targeted, vested-interest audiences, television as a teaching medium generally has not achieved the results or gained the practical credibility expected of it. Yet, television should be the most magical of all teachers. It has the capacity to invest fantasy with reality and turn reality into fantasy. It can heighten the mundane and charge ordinary situations with electricity.

Television, however, is not particularly well-suited to the direct, unedited presentation of real-time life. Although superficially it may seem to show events as they happen, these events are, in fact, significantly modified for presentation to the viewer—they are, in effect, "instructionally designed." For example, news stories are edited into their most salient aspects so that events always seem urgent and "fast breaking." The many hours spent waiting for developments during a hostage-taking incident are not reflected in the news coverage of the dramatic but very brief moments when the police intervene or when the hostages are released.

Even live sports events, such as football or baseball telecasts, are "manufactured" to an extent by television. There are teams of "designers" selecting, reorganizing and modifying reality to improve its presentation. Of course, the designers are acting mainly after the fact, but their intention is to create a program that goes beyond the perspective of a single viewer in the stands. No one spectator at a stadium could possibly see all of the aspects of a football game shown to television viewers. Without the benefit of instant replay, super slow motion, camera isolation and certainly the view from the blimp, the stadium-bound spectator sees only one reality of the game and a very limited reality at that. The television viewer sees a game which is "real" in many more respects since the viewer is able to see it from several perspectives. The television game has been designed to create this effect.

The dynamism of broadcast television and its ability to hold audiences, especially children, seems directly related to its capacity for presenting apparent immediacy. This immediacy is achieved by collapsing time and space to encapsule an incident which strikes a universal chord in the collective experience of the viewers. In doing so, television is not essentially different from any other creative endeavor, as it modifies the mundane, recreates it and represents it in a way which is at once both fictional and factual.

With very few exceptions, highly successful commercial television programs for children and adults have taken real-life situations into the arena of fantasy and created programming which seems to many viewers to be more real than reality itself.[1] An example of a highly popular show which understands and uses this fantasy/reality mix is *The Muppets.* The character "Miss Piggy" became so real that she made the cover of a prestigious national weekly as a candidate for the presidency; "Kermit the Frog" appeared on the 1981 Academy Award program. High-selling, commercial broadcast programs such as this are not burdened by the needs or objectives of instruction. They are, of course, carefully designed to be provocative and informative, but are rarely intended to do more than attract and hold the biggest audiences possible.

Instructional television, unlike commercial programming, fails in its purpose if it fails to teach, no matter how numerous its clients. The critical dilemma facing the instructional designer of interactive video is how to take advantage of the magical properties of television production and invest them with effective instructional capacity that does not impair the attractiveness of the programs. Mere transposition of current instructional design techniques from educational television will not suffice, for at least two very good reasons.

First, almost all previous design has been based on the "passive viewer," watching the program from beginning to end with very few, if any, interruptions. Almost all interaction takes place with the teacher, for whom the good program has become an extension of self and who thus becomes the mediator of the learning and interaction. Second, instructional designers for television have always developed a program with its broadcast function in mind. This has resulted in modified, controlled and conditioned design in terms of program length, scheduling, sequencing and use.

In interactive video, instructional design *is* the teacher. The design creates the scope, extent, nature and form of the interactive learning experience. Properly handled it replaces the teacher in the medium and itself becomes the mediator between the content of the audiovisual program and the learner. It encourages, reacts to, redirects or rewards performance. The content of the program will be enhanced and the user will learn to the extent that the instructional design is an effective teacher.

The virtual absence of broadcast time and placement constraints will free the designer from the enforced linearity of broadcast programs or non-interactive video tapes. This new freedom has major implications for

1. G. Gerbner and L. Gross, "Living with Television: The Violence Profile," *Journal of Communication* 2(26)(1976): 173-199.

the designer. Programming can be created for use at any time and for access at any point. In effect, an interactive video program can become as many different programs as there are users with different needs, entry points and learning patterns.

Such interactive programs offer the instructional designer speed and flexibility exceeding that of even the most elaborate printed or computer assisted programmed instructional materials. The effective designer must provide the interactive video system with a fully patterned format. The format must be complex enough to take advantage of the technology's capacity to become a personalized, electronic classroom, equipped with a masterly "teacher" in the guise of a totally responsive and highly flexible instructional design.

PRINCIPLES OF INSTRUCTIONAL DESIGN

Traditional principles of instructional design are well known. All apply in part to the special needs of interactive video, and will be discussed in turn.

Needs Identification

Even now, programs are designed, produced and sometimes delivered in the absence of any real need for them. Someone with a "good" idea who has the position or funding necessary to create a program determines that one shall be made and, quite often, blames the technology or the medium when the program fails to sell or is not utilized. If definition of need is important in educational television, it is absolutely critical in interactive video. The costs of production and mastering, and the special purpose nature of interactive video programs aimed at narrow audiences make needs identification the key to marketing and the basis of design strategy.

Techniques of needs identification in interactive programs may be markedly different from the traditional approaches applied to educational television. At least initially, interactive video users are likely to be more specialized than are the mass audiences in broadcasting. Their needs may be specifically associated with narrowly defined skills improvement or work and management techniques. Therefore, identification of instructional needs will be multifaceted. It will have elements reflecting the current availability of competing materials; the extent of market demand; the life of the program; the nature of the demands of the client group, including both buyers and users of the interactive system; and the overall placement of the interactive components in the total training program.

It may be necessary to establish a new and innovative approach to commercial and industrial needs identification. The market research techniques currently used by broadcast production houses might be adaptable to interactive video with respect to surveys, consultants, advisory boards and commissioned market analysis studies. But for maximum effectiveness, interactive video needs identification will probably be much narrower in scope and should focus first on the objectives of buyers who have sufficiently large numbers of trainees requiring personalized instruction in very specific subject areas. Hence needs identification techniques may be applied more to the organization's purposes rather than directly to the targeted user as has traditionally been the case.

This suggests the use of market analysts who are able to perceive and describe an organization's training or instructional objectives and then interpret these objectives in an interactive video format. Supported by target (individual user) analysts, the market analysts can prepare an assessment of organizational and learner needs. From this assessment, decisions to use interactive techniques and regarding the development of appropriate instructional designs can be made.

Analyzing Target Audience Characteristics

Analyzing target audiences will be a major task for instructional designers working with interactive systems. Tighter and more complete assessments of very narrow groups of individual users will be necessary to take full advantage of interactive video. Awareness of the specific characteristics of the users will be more valuable to the designer than the more general analyses provided for large-audience, linear programming. Therefore, the designer should seek information on individual differences, including characteristics such as cognitive style (the way an individual handles information—whether he or she prefers low or high complexity content, deductive or inductive approaches, print or pictorial information, etc.); levels of education; prior training; motivation for taking the courses offered; and ability to receive and process information.

Of course, this does not mean that instructional designers should design a program to one individual's personal characteristics. Rather, they should narrow down the range of mass characteristics, and closely identify the key traits, demands and learning strategies of a very small (by broadcast standards) but specifically interested target audience.

Not only must instructional designers recognize the special nature of the target audience, they must also note a critical element that often prevents excellent programming from reaching its target: the "gatekeeper," or the person who buys, orders, selects or utilizes the program and who must be convinced of its value if it is to have any opportunity to do its work.

Importance of Gatekeepers

Gatekeepers in instructional television can be elected officials, school board administrators, principals, vice principals, department heads and teachers. Sometimes they also include parents and community groups. Any one of these gatekeepers can block a program from its intended audience. Designers of interactive video for schools or institutional use will need to create programming objectives and design presentational strategies with these gatekeepers very much in mind.

Equally important are gatekeepers in the commercial, industrial and bureaucratic worlds. No one gets to use the programs without their approval, hence they should be included in the target audience analysis whenever a new project is under design.

Design of Instructional Objectives

Even in a very complex broadcast, such as a one hour program of *Sesame Street,* design of the instructional objectives is relatively simple. An educational goal is set for each segment of the show and a suggested means of achieving the goal is provided for scriptwriters. There may be up to ten educational objectives in a program. The designer's job is to ensure that they are clearly stated and that they are incorporated into produceable scripts. Each segment is linear (starts, has a middle, and ends) and cannot be accessed at multiple entry points or re-sequenced within segments. Almost all instructional programming for both children and adults is designed in essentially the same fashion.

Interactive video requires a very different approach and demands a more complex integration of attraction, motivation and educational content. Nevertheless, the principle of integrating instructional material with entertainment still applies.

Task Analysis

In the area of task analysis the increased flexibility of interactive video is a new factor for the designer. Unlike traditional methods which approach a task analysis within the rigid sequencing imposed by broadcasting, new approaches will be based on the almost instantaneous accessibility of any frame or sequence coupled with a dual audio system and advanced computer access for programmed instruction, testing and evaluation. Therefore, task analysis becomes the determining factor in setting up an interactive program and creating its limits.

Scripting

In major production houses there has always been a marked division of labor between the instructional designer and the scriptwriter of educational programming. In producing smaller, industrial programs both functions are often performed by one person. For interactive video, designing and writing are practically an integrated task. This is especially true when using the video disc, since much of the content might be presented in various graphics rather than exclusively in action sequences common to sequential film and video tape programs.

Because the instructional designer is heavily engaged in creating flow, branch and re-entry patterns within the interactive video program, he or she needs to work exceptionally closely with scriptwriters to produce a coherent and effective program. Consideration must be given to such design elements as integration of action sequences, static displays, special slow- or fast-motion passages, branching, testing, assessing and redirecting, and presentation of accessing instructions. The complexity of the task for the designer far exceeds any prior experiences, including programming for computer assisted instruction—a technology which possesses neither the flexibility nor the comprehensiveness of interactive video, and lacks much of television's appeal.

Therefore, the new wave of interactive video instructional designers will confront a new set of design questions, many of which are as yet unknown. Some, however, are already apparent, and these are discussed below.

SPECIFIC DESIGN ELEMENTS FOR INTERACTIVE VIDEO

The multidimensional nature of interactive video creates a number of key activities for the designer. Essentially, these can be broken down into an assessment of system capabilities; an analysis of strategies; development of modules; the creation of flow charts (storyboards); the integration of dynamic and static components; and the development of branching, sequencing and testing.

Assessment of System Capabilities

As the Corporation for Public Broadcasting's excellent technical report points out,[2] it is critical for instructional designers to clearly understand the nature of the interactive system which will carry the content. Such

---

2. *Videodisc: An Instructional Tool for the Hearing Impaired,* Technical Report no. 25 (Lincoln, NE: University of Nebraska—Lincoln, 1980).

questions as the need for motion; the most appropriate symbol system (digital, iconic, analog or combination); the degree of linear sequencing required versus the need for branching; and control of instruction (teacher initiated, student activated, or computer controlled) all need to be known and considered.

Video Disc Systems

The video disc is capable of combining almost every known visual medium and displaying it accompanied by either of two audio tracks, and then reordering the material in virtually limitless combinations. The designer's task at this stage of interactive video development is to select and organize the capacity of the system rather than find ways to test its limitations.

Designers should be aware of some limitations of video discs. For example, discs do not handle dense print well. Although an individual user close to the screen can read a reproduced page, it is difficult; at normal viewing distances or in small group settings it is impossible. Electronically generated print is handled very well, but is limited to eight or 10 lines of 20 to 30 characters per line for effective viewing. This means that a designer wishing to use large amounts of printed display must use either a significant number of frames or risk the target audience's being unable to read the display. Furthermore, single frame displays of print graphics are not accompanied by audio, which may create comprehension problems for a poor reader.

The difficulty in using blocks of readable print is also evident when incorporated with graphic designs, or when superimposed on other visuals. It is important to develop a storyboard, or at least a mock-up of print/visual integrations, to ensure both aesthetic value and comprehension.

Perhaps the most important factor in assessing the system capacity is the determination of the controller, or system activator. With video disc systems there are elements of the broadcast television system (largely teacher controlled); the programmed text (student controlled); and computer assisted instruction technology (computer controlled). Balancing and exploiting each of these levels or agents of control lies squarely within the domain of the designer. Clearly, there will be an enormous temptation to fully computerize programs developed for complete capacity discs attached to microcomputers. However, it is important to remember that not all gatekeepers will appreciate being outmanned by a micro, and most students will appreciate some control over the process. In order to make the program as close to truly interactive as current video disc technology permits, the essence of control probably needs to be a shared responsibility. This may be achieved by making the microcomputer program responsive to commands built into the disc with an additional override, or supplementary access, available to the user. Good balance should be the guiding principle. In any case, it is clear that the "teacher" inherent in programmed interactive video is the instructional designer, exercising control over the content and integration capacity of the interactive video tape or video disc.

Analysis of Strategies

The capacity of the interactive video system will to some extent determine the user strategies, especially as these apply to instructional design. For example, a basic system set up for home use will not offer easy integration of stills and motion on demand, although it can and does provide linear motion and still frames. Clearly, a lack of random access and integration capability would make all but rudimentary programmed instruction design very difficult to achieve.

Given access to programming, the designer gains much greater flexibility and can incorporate some branching, multiple choice testing and performance assessment/reaction. When the microcomputer is adapted to the system, full function becomes available so that the user's interactive control over the content (but not, of course, insertion of *new* content) is feasible.

In designing the instructional strategy, the simplest approach is the straightforward transfer of existing linear programming, for example, from tape to disc. The programs can be broken up into segments, and testing can be inserted. However, in this approach only a fraction of the disc's capacity will be used. Early development of instructional video discs followed this mode without either markedly enhancing their own effectiveness or illustrating the capacity of the system to teach interactively.

The transfer of existing programs presents little challenge to the initiative or creativity of the instructional designer and fails to take full advantage of the technology. Merely transferring existing programs should be regarded as, at best, a stopgap, and as potentially very harmful to the future of interactive video.

More interesting and worthwhile is the problem of integrating the various video and audio components to create interaction with the user. At this period in the development of instructional design, it would seem that strategies bent upon making the most of system capacity must begin from the ground up, rather than attempt the adaptation of existing materials.

Creation of Instructional Modules

Instructional modules refer to segments or sections of the interactive program, or the program itself as part of a package. For example, one instructional module can be a segment in print, another can be a segment of live action, and yet another can be a segment of questions.

In many ways, interactive video is a composite of all former audiovisual teaching systems. It can, conceivably, reproduce a book, a silent filmstrip, a slide/sound show, a video tape or film with or without sound. It follows that the instructional designer can create educational strategy, or instructional modules, for any of these teaching technologies. What is unique about interactive video is that it can integrate any one of these technologies with almost all of the others (sound must accompany motion). Even more exciting is the capacity to develop modules from modules.

Preplanning the branching systems is essential for good design of instructional modules and interactive programs as a whole. The techniques of computer assisted instruction are directly applicable, but a danger exists in making the system so mechanical and repetitive that its appeal and motivational values are reduced. Despite their remarkable flexibility, interactive systems must contain design elements that maintain interest in the content as well as the technology.

Creating Storyboards

Creating storyboards is an important means of ensuring coherent instructional design in educational television and film making. It is essential in the more complex design of interactive systems. Development of an effective and comprehensive storyboard is an absolute necessity in integrating the elements of instruction, branching, motivation, reinforcement and review which should be present in interactive programming.

The interactive storyboard will possess features not found in those created for regular television programming. One of the more simple examples of a storyboard is that designed by Bennion for the video disc (see Figure 5.1). This approach uses two visual and two audio columns. Visual 1 contains a description of the type of material displayed on the monitor. Visual 2 lists the content, placement and detail of captions overlaying the visuals noted in column 1. Each frame is numbered and instructions regarding the type of mode (still, motion, etc.) are provided. The two audio columns are representative of the two distinct audio tracks available on disc. Their content provides the same information as would a regular filming script. It is essential that the two audio tracks remain distinct, directly related to a frame-by-frame reference and properly aligned with their appropriate video segments.

More detailed storyboards are required when still and motion video modules are integrated into multiple choice branching, sequencing and testing strategies.

Figure 5.1 Sample Storyboard Page for Video Disc

File ____________

Page ______ of ______

Video	Captions	Audio #1	Audio #2
			1
			2
			3

Source: Adapted with permission from Junius L. Bennion, *Authoring Procedures for Interactive Videodisc Procedures—A Manual* (Provo, UT: Division of Instructional Research, Development, and Evaluation, Brigham Young University, 1976).

Integration of Still and Motion Units with Multiple Choice Branching

It is beyond the scope of this chapter to offer a detailed method of accessing the interactive system for branching, but the critical elements needed for an effective learning strategy can be identified.

It is critically important to recognize that the microcomputer is in control of presenting the material in the manner conceived by the instructional designer. The tendency to design for the microcomputer rather than for the student must be avoided. In effect, the microcomputer should be viewed as a means of accessing a coherent systematic display of the modules according to the performance of the user within the parameters of the program design.

Thus, the strategy for achieving full effectiveness of the interactive system is based upon using the microcomputer to call up displays, respond to input, branch to new modules, replay where necessary, test performance and assess progress. Naturally, all these processes must be programmed. This can be done at the start of the program or it can be fed separately to the microcomputer.

In a major undertaking such as a full function interactive program, there is a clear need for a programming specialist to translate the designer's video and audio components and branching or instructional strategies into computer programs. The instructional designer, therefore, should concentrate on the content and educational processes of the interactive video program while leaving much of the actual computer programming to the specialist. Nevertheless, a close watch must be maintained to ensure that the programmed instructions are consistent with the strategies and demands of the designer.

Sequencing and Testing

Perhaps the most compelling facet of interactive video for the educator is its capacity to test the user and assess his or her progress, then react to it, much in the style of a good teacher. If instructional designers can master the art and science of developing complete teaching units in specialized video packages, they might well change both the structure and quality of education, particularly at the college and university levels.

Interactive video may develop home study and self-instruction patterns that would eliminate the necessity for much basic course work in high school and college programs. Interactive video could become the main instructional method of reaching the majority of part time and continuing education students if it is designed to ensure effective course completion and accurate assessments. Programs would need to be designed to provide content and test performance, and to use or modify many of the principles and practices of adult learning. This practice of applying or adapting these principles would be most apparent in the structuring of instructional sequences and performance evaluation. Accordingly, the instructional designer would need to become aware of critical adult learning principles and their application within the context of interactive video.

Applying Learning Principles to Interactive Video Design

No one has truly isolated the critical principles involved in learning. They are thought to be an integration of motivation, intellectual development, intelligence, experience, practice, reinforcement and all sorts of factors relating to personality, drive, need and self-esteem. This chapter is, therefore, clearly not the place for an "in-depth" analysis of learning styles and principles of instruction. But it is important to note that good teachers over the years have discovered many practical approaches to gaining and maintaining the learner's attention in order to make learning both easy and effective.

Essentially, these practices involve motivation (the desire to learn), attention management, organization and coherence of material presentation, practice and review, reinforcement, and recognition of success, either through test scores or the granting of certificates or diplomas. There are, of course, a great many other techniques and variations, but those noted above seem to survive the tests of time and systems.

The question confronting the designer of interactive video is how these techniques, previously promoted and maintained by the human teacher, might be incorporated into the program.

Motivation

Initially, there is a built-in motivation in interactive video in that it is a new and exciting technology. This "toy" factor will pass once familiarity is established, and interactive video will have to compete with all other teaching technologies. Its success then will most likely depend on how relevant the materials are to the needs of the learner, especially the self-motivated or industrially primed user (one who is required by management to grasp the content and/or skills contained in the program).

Thus, the most important aspect of the designer's work is to incorporate content and techniques into the instructional design of the program that are totally relevant to the needs of the user. At the very least the design must reflect the essential needs of the vast majority of the users or it will lose credibility in the eyes of both gatekeeper and student. Highly relevant, truly needed content will cover a multitude of faults—even, at times, a dull or poorly produced program. Of course, this is not to suggest that such faults are acceptable or can even be condoned. However, it does illustrate how important relevance of content to need is in the designer's conceptualization of the program.

Motivation is significantly enhanced if the content is well-produced, interesting and not burdened with too much detail. Motivation is reduced by obscurity of purpose, lack of reinforcement, poor pacing, inappropriate levels of difficulty and incompetent production techniques. Therefore, a very clear understanding of several factors is essential to successful design. These are not only the needs, learning styles, purposes and intellectual characteristics of the learner, but also how these elements relate to the user's expectations, should he or she successfully work through the program.

Management of Attention

Management of attention in a television program is perhaps best illustrated by the preschool children's program, *Sesame Street.* Common wisdom has it that a child's attention span is extremely short, perhaps on the order of 20 to 30 seconds. Yet *Sesame Street* can hold a three-year-old almost immobile for an hour. It achieves this apparent miracle by managing the pacing and production content of its magazine show so that children are in turn excited, relaxed, entranced and interested by widely varied segments. If all the fast-paced segments were played one after the other the children would probably become inattentive. If the slower paced parts were similarly handled they would change channels. It is likely that adults would react similarly. The principle underlying *Sesame Street* is variety, not just in content but in production style, to take advantage of the viewer's apparent need for successive re-stimulation.

The lesson for the interactive video designer is to remember that the content will be displayed on a television screen and will have the characteristics of a television program. It needs to be designed with variable pacing in mind. The program should not be expected to succeed purely on the basis of relevant content. Timing, variation of modules, rest periods containing relatively low demands in content or performance and an awareness of the need to interest the imagination as well as feed the intellect should be essential parts of an instructional designer's strategy.

Poorly organized or incoherent presentation of content is dangerous to any instructional effort. It is exceptionally so in interactive video, partly because it denies the medium its full capability, but also because it will frustrate a user who *expects* a programmed instruction system to be well designed and very efficiently organized. If the presentation of materials lacks coherence or confuses the learner, the essential branching and assessment functions will be jeopardized.

Practice and Review

A major drawback for linear educational television used at home or in the classroom is its inability to provide credible and effective practice and review. It can and does offer reprises of segments, sometimes as direct repetitions and often as reviews in varying formats. But it cannot offer the hands-on practice that is available to the interactive video user. Practice and review are traditional cornerstones of educational theory and application. Therefore, the designer should go to substantial lengths to incorporate legitimate practice in applying, repeating, remembering or interpreting the content. Repetition without interactive work-throughs short changes both the medium and the user.

Reinforcement

The capability to directly and personally reinforce the user (both positively and negatively) is a very strong asset which must be exploited by the designer. Much of this reinforcement will occur through success in tests and completion of program segments. But interactive video can also be programmed to take a "personal" interest in the progress of the learner. Instead of an automatic and impersonal return to reteach, or clicking forward to the next segment, the disc or tape can be made to react, perhaps in the manner of some Las Vegas slot machines which groan when one loses and rejoice when the jackpot is hit. Of course, one might expect a little more sincerity from a supportive interactive program.

The concept that reinforcement aids motivation is a well-established educational principle. It is not as certain that it actually increases learning. But interactive video should not be all work and no play. Integration of the viewer's success on tests with some built-in specific reinforcers, touched with humor, should be included in the designer's tactical bag. Another valuable tool is an evaluation plan so that the ultimate effectiveness of the program can be assessed and techniques and strategies improved.

EVALUATION PLANNING

Almost every request for proposals issued by granting agencies or industrial and educational television users requires an evaluation plan to assess the results obtained in the project. Too often, these plans produce useless data, and they are sometimes regarded as little more than chores necessary for securing the funding. Typically, such plans call for an analysis of learning and attention variables. They are beset by a great many technical research difficulties inherent in presenting the materials to large groups simultaneously. Distraction, control of prior learning, the experimental conditions themselves and the fact that the participants tend to see themselves as critics rather than viewers, result in data of very doubtful quality. Interactive video has a built-in capacity for almost pure data gathering that can be easily exploited by the designer working in conjunction with the microcomputer programmer. This is especially true when a print-out of performances can be obtained.

Designing an Evaluation Plan into the Program

The implication is that the evaluation plan can be designed directly into the video disc or tape by recording and storing it either within the microcomputer for subsequent display or in the form of a print-out for subsequent analysis. This plan will provide direct, in-use, individual data for every user's performance. The variables might include some or all of the following:

a) Total time for completion of the program.
b) Number and nature of branches needed.
c) Number and nature of segments not used.
d) Number, nature, and degree of difficulty encountered in segments requiring repetition or branched reteaching.
e) Total and partial scores on all tests.
f) Item analyses of all multiple choice test questions.
g) Reactions to reinforcers.
h) Evaluation by the user of the total program after completion in terms of:
   1) content relevance
   2) ease of understanding
   3) effectiveness of segments
   4) difficulties encountered
   5) interest levels
   6) attention holding

In effect, this procedure makes the program its own test instrument and provides a much purer test of its ultimate effectiveness. This capability eliminates the subjective bias of the traditional observational, self-report and post performance assessments.

The data can be collected, stored and analyzed in the microcomputer. Individual performances can be compared to preset norms and subsequently incorporated into the norm if so desired. Also, the video player equipped with a microcomputer can be programmed to store individual data, combine the responses of users and perform most of the data analysis functions normally required in program evaluations. Should data be required from numbers of users, a system of print-outs or retrieval of stored data would be necessary. In either case, programming an evaluation plan directly into the video disc or through the microcomputer will provide a much cheaper and certainly much more accurate assessment of the interactive program's performance.

It is unlikely that the instructional designer will be an expert in evaluation; the task requires a research specialist with whom the designer must work closely. Evaluators have a tendency to design instruments that are excellent in terms of research, analysis and coding, but which sometimes are not totally related to the content and specifics of the program. They need to understand the content, purpose, strategy, limitations and expectations of the program as it *was* designed not as it might have been. Close liaison is essential for meaningful evaluation.

Once committed to mastering, program revision is a costly undertaking. After-the-fact evaluation can be very useful in developing new projects, but is not very valuable in revising in-use interactive video programs—unless a new edition is planned involving remastering.

Special Concerns for Video Disc Evaluation

The type of research that can affect the development of new materials before they are mastered is formative research, which tests segments and techniques as they are designed and modifies them prior to mastering. The problem in adapting a formative research plan to an interactive video disc lies in the fact that the completed disc cannot be changed and that the full functioning of the disc cannot be achieved until the tape or film master is transferred to it. This "Catch 22" situation will probably force costly pilot evaluation of major projects involving video discs made expressly for formative research projects. In any case, the instructional designer must stay very close to the action since defense of the original, and incorporation of relevant findings, will be his or her responsibility.

CONCLUSION

Much is yet to be learned about the nature, tactics, strategy and assessment of the instructional design of interactive video. But it is certain that the designer is to the interactive video system what the producer/director is to network television. The designer is the critical element determining the style, efficiency, quality and ultimate success of the enterprise. Like the producer/director, the designer is dependent on the project team. Should too many roles—especially those of segment writer, programmer, producer/director and evaluator—be taken over by the instructional designer, the project will be weakened. In effect the designer must be aware of all the functions performed by others but must also allow each member of the team to exercise the special expertise his or her role demands.

The essential strategy of the designer is the integration of the whole from all of its parts. The best evaluation of the design itself will come from the knowledge that appropriate use of the system's capacity has been achieved and that users succeed in their efforts to interact with, and learn from, this new, exciting and still unexploited medium.

6

Producing Interactive Video Programs

by Patrick McEntee

Producing an interactive video program is more complicated and time-consuming than producing traditional linear video programs. The complexity of the production process requires the producer to coordinate a number of unique factors. These elements include: selecting and managing the production team; setting up workable guidelines for the production; selecting a production facility; and finally, planning and controlling each step of production. In order to clarify the issues you will be confronted with in producing an interactive program, we will review each of these points in some detail. Let's begin with the production team.

SELECTING AND MANAGING THE PRODUCTION TEAM

Although most members of the production team will perform the same functions as they would for a linear program, two new members assume much greater responsibility: the instructional designer and the computer programmer. These two positions are generally not even included in linear production. The instructional designer is responsible for providing clearly defined objectives, a flow chart (approved by the computer programmer) and an evaluation plan (approved by the content expert). The computer programmer is responsible for writing computer codes that create the interactive dialogue, and then programming the codes on the tape or disc.

Frequently, the people assuming these roles will be unfamiliar with the requirements of video production. However, their participation and co-operation are critical to maintain continuity from the design phase through production. When unforeseen problems arise during production, they can help make needed changes to eliminate expensive delays or revisions.

If a program requires certification of the learner's performance, an applied researcher should be hired to validate the program. The client may specify these performance statistics as essential, and view their provision as part of the production process. Sometimes the learning strategy approved in the design phase falls short during the formative evaluation, forcing changes in production which delay the schedule and increase costs. To avoid this, plan production in detail, and confirm that the authoring team has accurately completed the objectives, flow chart, script and evaluation plan before scheduling production.

Professional Planning and Control Techniques

The complexity of an interactive video program can be effectively managed by using professional planning and control techniques. The project manager or producer should define step-by-step guidelines for performance, evaluation and revision. Appendix 6A (at the end of this chapter) presents a plan with 40 steps that can be used to monitor the progress of an interactive production. It will be referred to as the steps of production are discussed in detail.

Naturally, the evaluation and revision process measures a team's commitment to quality program development. As production begins, undetected errors made in the authoring stage may force modification of production plans. Therefore, the project manager should generate options in the plan. A detailed familiarity with the objectives, the learning strategy, the evaluation plan and a clear set of goals specified for each working group in the project team will help the project manager develop these options. Knowledge of all other parts of the program planning process allows the project manager to confidently create a cooperative atmosphere for team work. If the worst happens, and it becomes clear during production that the program must be modified extensively, then the team can work together to resolve the problems.

Integrating Computer Techniques

Since interactive programs are made possible through the use of microprocessor technology, program structure and information are dependent on computer logic and conventions. As a result, a producer who understands how to merge the computer with television can tailor the content to add style and form to an otherwise dull program. The project manager, the instructional designer, the producer, the director and the computer programmer should try to make the technology transparent and accessible to the learner.

An interactive program should create an artificial time for the learner during which curiosity sets the pace. The program should proceed at the learning rate that the viewer needs, and allow searching based on references that the viewer chooses. Freedom to search in different ways helps to personalize an interactive program.

Production Guidelines

Production guidelines are still dependent on market competition. Currently, there are several tape and disc formats, no common standards for video quality and little coordination among computer codes. Gradually, manufacturers are developing informed support staffs, but the video production community (including creative houses and facilities) remains for the most part unaware of interactive software. As a result, the project manager should clearly define guidelines to overcome the lack of familiarity with interactive production. The project manager should also set forth procedures for printing a guide or workbook. If program content is changed, then the workbook must be revised to match the new data, and audio and video segments. So, plan the workbook, make a mock-up and test it in both the formative and summative evaluation steps, but do not print it until the entire program has been tested, revised and approved.

SELECTING PREMASTERING FACILITIES

After selecting the key members of the production team, the next major decision is selecting a production premastering facility. Premastering is simply the process of creating a master video tape that will be duplicated on video cassettes for distribution, or sent to a manufacturer for mastering and replication onto video disc.

Before selecting a facility for production, it is necessary to develop a set of criteria for each project based on the specific media and the type of program. One of the biggest problems in interactive video production is assembling still frames during the premastering phase.

Film-based vs. Tape-based Production

Whether you are working toward a final release on video tape or video disc, still frames such as menus, indexes, question prompts or compressions of longer sequences are an essential part of the program. On video tape these still frames are recorded for a fixed duration, but on video disc, still frames should be recorded for one frame only unless simultaneous narration is necessary.

It is necessary to verify that the facility has the capability to produce high quality still frames, as well as motion sequences, audio and computer codes. Frame-accurate editing capability is essential for video disc programs and preferable for video tape programs.

Still frames can be created and edited on either film or video tape. Table 6.1 provides a comparison of film and video tape characteristics that you can use to evaluate a facility.

As you enter a post-production facility, keep the basic characteristics of film-based versus video-based production in mind. You may want to make a short checklist, such as the one given in Table 6.2, and check off what each facility has to offer. This facility check comes before Step 1 of the production process list (Appendix 6A), and in fact could even begin in the authoring phase.

Table 6.1: Film and Video Tape Characteristics

Characteristic	Film-based	Video-based
Production Method	Separate production of slides, filmstrips, and video tapes	Total capability with computer graphics if desired
Time Required	Time lost for separate production and viewing	No time lost except for the original art required
Generation Loss	Each separate production requires a generation loss or several	Eliminates generation loss of film-based media
Revisions	Complete artwork, photo, slide, film-strip revision	Revisions made in "real time" during formative evaluation
Acquisition Costs	Some equipment already exists typically in studios —relatively low cost	Relatively high costs of hardware acquisition
Production Costs	High recurring costs because of artist time, photography, and film processing	Eliminates many recurring costs which will amortize cost of system rapidly

Source: Adapted with permission from H. Deway Kribs, "Authoring Techniques for Interactive Videodisc Systems" (Instructional Science and Development, Inc., 1981.)

Table 6.2: Facility Checklist

Film-based		Video-based	
Type composer	_____	Character generator	_____
Layout & mechanical	_____	Video compressor	_____
Optical printer	_____	Advanced switcher	_____
Animation stand	_____	Digital graphics	_____
Genigraphics	_____	Still frame storage	_____
Color correction	_____	Color correction	_____
16mm editing	_____	¾″ video tape (time code)	_____
35mm editing	_____	1″C video tape (time code)	_____
Qualified editor	_____	Qualified editor	_____
Audio editing	_____	Audio editing	_____
Audio mixing	_____	Audio mixing	_____
Detailed edit lists	_____	Computer file editor	_____
Rank cintel	_____	Rank cintel	_____
Screening facility	_____	Screening facility	_____

As you check each of these capabilities remember that each service costs money. If you are under tight financial constraints, yet feel that this program should be produced on video disc, investigate film production carefully. Your production staff may have considerable experience in shooting and editing film, and that experience could be an advantage in assembling still frames. However, all interactive programs are used in video format and, therefore, all materials produced in film will ultimately be transferred to video tape for mastering.

Whether you have decided to assemble the program on film or video tape you must know whether the final program you are making will be a video disc or video tape. Generally, this decision is made in a needs analysis before investigating facilities, but your visits may uncover opportunities to save money. Also, the program may not require all of the potential interactivity of a video disc to accomplish its objective.

Video Tape vs. Video Disc

There are several trade-offs in deciding between tape and the optical video disc that will determine the style, quality and flexibility of your program. Basically, questions about tapes versus disc fall into four categories: access time, complexity, permanence and cost. In the rush to make a novel video disc, many poorly planned and executed video discs were pressed, which later became "just experiments." Clarify your exact reasons for making a video disc rather than a video tape.

Video tapes can be updated with little effort; video discs can only be updated with a data overlay. Video tapes can be sent out to a user almost immediately; video discs must be mastered (a cycle that takes seven to 10 weeks after premastering.) However, video discs do several things that video tapes cannot. The following series of questions will give you an indication of whether a video tape or a video disc is more appropriate for the production of your program.[1]

- How many copies of this program will you make?
  10-100: video tape may be better
  over 1000: video disc becomes cost-effective

- When do you need the program for distribution?
  If you need it today, use tape
  If you have seven to 10 weeks, consider disc mastering

- What is the shelf-life of this program?
  Quick turnover programs belong in tape
  Disc is more appropriate for classic programs

- How many times will each copy of the program be played?
  Tape deteriorates with every play
  Disc is a nearly permanent resource

- How much branching does your program require?
  Video tape programs tend to be more linear owing to cue and search time

1. Adapted with permission from "Needs Analysis" (New York: Sony Video Products Co., 1981).

Complex branching requires the search speed of video disc

- What is the acceptable maximum cue time to reach the next module of the program?
  Video tape often requires a workbook to fill the cue time
  Video disc carries a larger percentage of print, and prompts the viewer with still frames

- Will your program be viewed individually or in a group?
  Video tapes tend to be better suited to group viewing
  Video discs tailor the program to an individual's needs

- Does your program require a large number of still frames?
  Still frames play in real time on video tape
  Video discs can store one still frame in 1/30th of a second

Some serious thinking on these questions should prepare you to answer the detailed questions that any premastering facility will ask you before they submit a bid.

Technical Quality

Whether you choose video tape or video disc, there are several questions that you will have for the premastering facility about their capabilities and costs. When screening a plant for its ability to produce an interactive program you must verify that the staff is conscientious and concerned about the technical quality of their product. You can easily watch for form and style but technical errors can remain hidden in the master tape until after it is duplicated. Aside from the general orderliness of the facility, do people seem at ease and relaxed in their work or is there an atmosphere of tension? A well organized, comfortable environment is crucial to successful production; tension breeds errors and hinders open discussion.

If the director comes with you to check the facility, a longer and more technical discussion could be valuable. That conversation should take place with the facility's production manager and engineer. The discussion should focus on these four issues: color balancing, still-frame composition, the character generator and computer programming.

Color Balancing

Color balancing is simply matching the color levels between two scenes. In linear video production, the technique is known as edge matching. The video operator plays down the edit file until two scenes next to each other are so noticeably out of color symmetry that the director asks for balancing. At that point it is necessary to rebalance the color electronically.

In an interactive program, in which the viewer can branch to several places, each possibility must be checked for color balance. Documenting each time the camera is balanced on location, plus the precise alignment of playback tapes to color bars before editing can minimize the overall error. Don't go overboard by asking the video engineer to check each color-balanced frame with every other frame. A quality facility with thorough set-up procedures that are uniformly observed by the staff will check the symmetry in the major program paths for color continuity. These set-up standards should eliminate most of the problem.

Still Frame Composition

The layout of still frames takes time, whether on film or tape. Either medium offers the opportunity to modify or correct mock-up after formative evaluation. However, revisions will be substantially more expensive in film-based production. Also, you may have to wait a day or more for the results of your film, since it must be processed. This should not delay your edit. Remember, *interactive programs can be assembled in almost any order* since they do not depend on a linear narrative structure.

The Character Generator

In video-based production, the character generator is a producer's right arm. Interactive video programs often have *more than 10 times the amount of text that linear programs have* and, for a complex video disc, that figure can climb much higher. Try to find a facility that has a character generator with at least two attractive fonts, in different sizes. If you are limited in the selection of font styles and sizes, you may want to focus on coordinating the question frames from a workbook. Adding color to the text will enable you to have two or more "voices" as long as these conform to an editorial style the viewer can understand.

Laying out text and pictures in still frames will be addressed later in this chapter, but one more comment at this point could improve your programs. The video compressor allows you to shrink one frame of video and place it within another frame of video. Thus, it allows the producer to present pictures and text in the same frame (similar to a magazine layout). This invention of high-end news and sports production, combined with the character generator, can give your productions added polish.

On the ABC/NEA SCHOOLDISC* team for instance, we used the compressor to shrink pictures so that we could place Dewey Decimal numbers over black in the upper right hand corner of still frames. As a result, children could find a library reference as they watched.

Computer Programming

The final performance criteria to check when researching a facility is the area that you are likely to know least about—data. Better known to the video world as computer codes, *data are the pieces of information that cue the microprocessor controlling the interactive branching.* Currently, the manufacturers of video discs will code the master video tape at the mastering plant, but that will very likely change in the future. Each video production facility will develop the capability to model and insert the interactive data on the master tape.

At present, however, don't be surprised if the production facility is totally unaware of data. You should be prepared to supplement the facility's staff with your own computer programmer, who has been with your team since authoring. The major point to remember is that you must leave the premastering session with a list of program commands and operands (frame numbers) that accurately represent the computer programmer's design for realizing the instructional flow chart.

In some video tape-based systems it is possible to program the master tape immediately, and evaluate the interactivity. For some video disc programs it is valuable to simulate the motivational segments on video tape prior to the check disc pressing. Pioneer Corp. offers a check disc option at a nominal cost to a customer for confirming computer codes. 3M offers cue insertion confirmation at several sites around the country. Sony Corp. provides customers an opportunity to simulate video discs at major production centers.

The production facility should modify your master tape at no cost to correct any errors resulting from its mistakes. However, if you make mistakes or your instructions were unclear, then you are responsible for the cost of revisions.

---

*The ABC/NEA SCHOOLDISC is an electronic curriculum designed for fourth-, fifth- and sixth-grade instruction. Content for the project is supplied by ABC News, Sports, and Wide World of Learning. Interactive software for the SCHOOLDISC was created by ISO Communications, Inc., and is available from ISO. The design of the program was granted the approval of the National Education Association.

Editors

For the most part, editors will understand the loading of data better than the rest of the staff because the computer edit decision list has become a standard tool in video post-production. This decision list of edits will drastically expand in an interactive program. The instructional designer and the computer programmer will often negotiate late changes with the director and the editor. So support your local editor—the producer must have the editor's cooperation.

With these guidelines in mind for selecting a premastering facility, we will review the production process, referring frequently to the 40-step plan given in Appendix 6A.

PRODUCTION PROCESS

A simple summary of the production process contains three major phases including preparation, shooting and premastering. Several pre-production meetings will be required to form a detailed production schedule and to delegate tasks. In the early meetings it may be helpful to have the instructional designer, computer programmer and writer attend for the purpose of continuity. In order to begin shooting, the format, script, graphics, motion sequence and narration must be confirmed and approved. The computer codes should be planned according to the flow chart, although this may change during shooting and editing.

Shooting, for the most part, is the same as it would be for linear video production. (However, there are a few tricks to remember, which are explained in the motion-shooting section.) Also, more attention is required as you shoot your still frames since so many things can go wrong (film versus video aspect ratio, unrelated layout, poor composition, illogical flow, etc.).

Documentation

Documentation means record-keeping. Keeping logs with accurate time code notes during production is very important, and often neglected. You will experience tremendous anxiety if you begin interactive post-production without accurate logs. The production assistant should be held accountable to provide all the information that your director, computer programmer and editor need in order to edit and assemble the program.

Provide several project and logging forms, with clear instructions regarding how they should be completed. Remind the production assistant that you will want to review the shooting process using these forms. Then spot check the log periodically during production to make sure that it is accurate.

Assembly

If preparation and shooting were well organized, the assembly of an interactive program will be straightforward, though time-consuming. If each segment has been thoroughly storyboarded there is little creativity in the editing of a motion segment, since it is often fixed in design.

The overall architecture and positioning of the segments in modules (discussed in Chapter 5) of the interactive program are entirely predetermined and are not changed unless there is a problem. At the start of the first production meeting emphasize the importance of documentation, and the publishing-like nature of the premastering assembly process. Even with these reminders, staff members will probably follow their linear programming habits so continually make sure that the documentation is accurate.

The First Production Meeting—Storming

Your first interactive production meeting may seem a bit disorganized. Actually, things may be going according to your agenda, but the sheer number of people (10 to 25 would not be unusual) increases the possibility of confusion. Even a simple interactive video tape may bring 10 people together for the first meeting.

As you introduce people to one another, subdivide the group into working teams. Generally, the character generator operator, director of photography and any artists should work through the director on creating a visual format that will increase the continuity between the text frames and the motion segments. The instructional designer, writer, producer and computer programmer should work together since they have seen the project start and develop. They may feel they know more about what should happen than the production group. The director will need to work as a bridge between these two groups.

The director should become the key player at this point, but if this is the director's first interactive program, he or she may be reluctant to assume a leadership role. If this happens, the project manager should start the ball rolling (Step 1) by outlining the goals for the project and presenting the PERT (Program Evaluation Review Technique) chart. The manager should explain the worse case delay situation or critical path, and then explain what each person is accountable for. The producer should explain the production budget and give creative direction, during which the group can ask questions. Next, the instructional designer should read the objectives of each module of the program and describe the strategies to achieve verified learning.

Generally, the flow chart and computer program are the most confusing parts of interactive production for persons unfamiliar with random-access software. Therefore, taking the time to explain the learning goals and strategy to the production team will result in better performance and more commitment from each artist. The computer programmer should explain the predicted flow of the program to the production team in simple language.

After reviewing the plans and strategies the program uses to support the learning process, the group should be more comfortable with the flow chart. Creative suggestions should be gathered at this point to avoid costly revisions and changes after production begins.

The First Production Meeting—Forming

Next, ask the writer to hand out the scripts. It may be helpful to read the script out loud. Even in linear programs reading the script provides a public critique of the language and the accompanying visualization. Reading the script out loud also helps to demystify the segmented structure and the logic flow of the production. You will sense if the meeting is proceeding well by the general level of participation. Each person will speak out in regard to his or her own specialty and the director will begin assuming control.

The First Production Meeting—Norming

By now the production team should be "seeing" the program, and criticizing the elements that don't work. As you read and reread the script together, remind the entire team that scripts are usually read in linear form. Consult the computer programmer to follow the notes in the left margin, and lead the group along the special program paths available for remediation and extra information.

Try to test the logic design from the different points of view in the group. For example, what do you offer slow learners or fast learners? Does the "quick study" get bored by being stuck in the mainline presentation for too long? One of the major advantages of using an interactive program is that it reduces learning time; use that advantage whenever possible.

The First Production Meeting—Conforming

At the beginning of the first production meeting people may be argumentative and unsure of the process and the script. The first meeting may expose inconsistencies in content that the instructional designer and the content expert may have overlooked. The director and artists may also spot bugs in the computer programmer's design. Sometimes flow charts appear so neat that they are deceiving—the cool logic of boxes and triangles may seem difficult to change, but generally they can easily be reorganized. Remember, the flow chart, sometimes known as the program map, is just a tool.

Encourage the computer programmer to work with both the instructional designer and the writer to incorporate the production team's suggestions and make revisions. It may be difficult for the instructional designer to analyze the implications for the learning strategy as changes are suggested. The designer may have to review the changes at another meeting, since many of the changes will need to be corroborated with the content expert. Any revisions will be smoothed by the writer after they are approved by the content expert, instructional designer and producer.

By the end of the first or second meeting the production team should have an idea of how much work is involved in the project. The producer or project manager should check the provisions made in the budget, and make any additional funding requests in writing before recording begins (Step 2). Before leaving the meeting, the computer programmer should create a preliminary program of control codes.

Gathering Materials and Recording

If you have been working with video tape as your production medium, you will not have to change your method much for shooting the motion segments of the program (Steps 2-21). The only consistent problem that the director and producer will have is image jitter. Each television frame is constructed from two fields of video. As a result, if objects are moving quickly in the video frame, then they appear blurred in still frame mode. One way to avoid this problem is by shooting detailed motion sequences on film.

Composition

The composition of an interactive program suggests an order to the viewer that is logical and offers a set of rules or conventions to manipulate the program. Therefore, the director's attitude toward the composition and layout of the interactive program is most important. We recommend using a director of photography to add further unity to the overall design. The director and the director of photography should agree on the thematic visual and spatial design of both the motion segments and the still frames (Steps 1, 7, 9, 11).

Sometimes, the director will require the support of an artist to prepare detailed storyboard revisions in restructuring the composition. If the director and producer fail to organize the visuals logically to establish continuity, the audience may react by focusing on the wrong information, or by interpreting cues inaccurately. As a result, the learner becomes lost in the program. Composing the video segments and the module architecture is not just pictorial packaging, but a method of creating a continuous interactive dialogue.

Whether you are working on a motion sequence, slow motion, video compression or a still frame, traditional composition principles still apply. However, the microprocessor and the interactivity that connects all of the production media operates as a dynamic composition.

It is important to keep in mind that the viewer can only see what is happening within the television frame. Today's televisions cannot reproduce stereoscopic depth for the viewer, only representations of reality. This artificial reality is even more noticeable in video disc where the still frames have no sound, and can be studied individually and in detail. The net result is that the overall graphic unity plays a much greater role by adding to the continuity or distracting the viewer.

Some of the still frame sequences may have to be shot on location, during the video production process (Step 11). Others are produced from the final content of the linear segments. Motion segments offer the chance to amuse the viewer visually, to motivate with audio and to build theme by theme. Keep in mind what video and data overlays are scripted for each scene, when shooting all motion segments.

Graphics and Titles

Graphics and titles establish the style and atmosphere of the program. Titles must be consistent sometimes and inconsistent at other times depending upon the pattern of logic they support. The typeface, size, color and background can be used to reinforce sequential and topical relationships. The size, placement and style of the text determine how effectively and in what way the viewer will read the content. Depending on the situation, effective programs can use 25 to 40 characters per line, and from eight to 18 lines of textual information.

The production assistant (at Steps 2, 4, 7, 9, 11 and 16) should monitor the text for accuracy and comprehension. The director should focus on using the still frame effectively for text storage. Graphics can be generated with any traditional media from film or video production, and from new data forms like digital graphics, three dimensional models and simulations. (Note: if you are using still frames in sequence they can be viewed in order at different rates.)

Media Transfer Rules

These rules correspond to Steps 2 through 21. If some (or all) of your source material is not on video tape, then you must take into account the following considerations:

1) *Aspect Ratios.* The relative dimensions of the viewing screen (horizontally and vertically) are referred to as the aspect ratio. Television has a 3:4 aspect ratio (3 units vertical for 4 units horizontal). Media differ in their aspect ratios. For example, 2 x 2-inch slides have an aspect ratio which does not match a video aspect ratio at all; 35mm slides transferred to video must be cropped slightly in width; 16mm film has a slightly different aspect ratio than video.

2) *Safe Title, Safe Action.* Television receivers and/or monitors do not show the entire video frame. Therefore, it is important to keep cues and titles away from the frame. The exact position of safe title area is displayed in most studios on a separate monitor. It is essential that information appearing in the program should be placed within safe title boundaries. Safe title is approximately 63% of the entire frame; safe action is 80% of the entire frame.

3) *Resolution.* Film is currently capable of higher resolution than video tape. Very small letters, numbers and details may be legible on film, but not on video tape. Color and contrast will also influence the final image in this respect.

4) *Transferring film to video tape.* Film passes through the projection gate at 24 frames per second, whereas video tape runs at 30 frames per second. Film to tape transfer equipment, or telecine, will automatically compensate for this discrepancy. The name for this compensation process is called "3-2 pulldown."

Each of these rules should be clear to the director and monitored carefully in Steps 2 through 21 as the visuals are laid out and recorded.

ASSEMBLY AND PREMASTERING

When editing begins, motion segments will be created almost exactly as they are in the linear production process. The rules of quality and continuity will apply for camera changes, transitions, duration and clear video and audio signals. Only the frames that cue the viewer and cause the program to branch (menu, prompt and question frames) change the editing process within a segment (Steps 22, 23, 24). The architecture of the motion segments, still frames and repeat plays is organized for the program assembly according to the flow chart provided by the computer programmer (Step 21). Several things should be kept in mind by the director during the editing process.

1) Video levels should be carefully monitored.

2) If portions of the program are to be accessed independently, good entry and exit points must be provided. Avoid dissolves, use cuts or quick fades to smooth segmentation.

3) In laying out the index of the program, consideration should be given to minimizing search time to access extra material. The computer programmer should have analyzed this earlier, but confirmation during assembly is important to catch errors.

4) Still frames can be placed in a bank of frames to facilitate editing. Or, they can be positioned logically on the disc or tape to minimize search times.

Premastering Format

In film, an editor and the assistants are responsible for the mechanics of the editing process; the director reviews their work. In premastering, the director runs the assembly with the help of the editor who is responsible for creating an electronic decision list. Every piece of video information must appear on the edit decision list.

Once the entire video and audio master is edited, dubs are made on 3/4-inch video tape and screened. Improvements are made after a review, if possible. After the content expert, instructional designer, and producer have checked the tape, the director will send the master tape to the mastering plant so that a check disc can be pressed. (A check disc is not always necessary. Sony, for example, has a simulation procedure that eliminates the need for a check disc.) Meanwhile, the instructional designer and the computer programmer carry out the formative evaluation on tape. When the check disc returns (Step 29) the director and the instructional designer evaluate it for accuracy. Then, the computer programmer debugs all of the computer code to support the formative evaluation at Step 29.

Computer Control Codes—Data

Control codes are the actual computer commands that are responsible for making the program play as planned (i.e., respond to viewer inputs, play in different modes, repeat). A final check of the program logic should be made to ensure that the program does what it is intended to do in the script, the flow chart, the shooting script, the detailed logic diagram and in the control instruction. Few productions follow the script 100% of the time so changes in the production and editing phases may cause revisions in the computer codes.

In each system the commands that the computer programmer writes are short statements like run, play, step, store, register. These commands are written as instructions with frame numbers as references (operands). Possibly, a short description of the actual machine function could appear at the right side of the computer programmer's notes. Other instructions could include what is to be seen, how much time will be allowed to view each segment or what mode to view the segment in.

The variety of preliminary instructions are like words in a sentence. The computer programmer will later translate control code into a detailed logic diagram. (Steps 29, 35, 38.) The computer programmer can instruct the microprocessor to begin executing the instructions from any line and it will execute the steps of the program sequentially beginning at that point. Every microprocessor has a fixed capacity, and each control code takes some space to store. Therefore, only a certain number of instructions can be written within the capacity of any microcomputer. The computer programmer's main objective is to use the data capacity as efficiently as possible. Revisions in the computer program are common; in fact, the entire programming process is a series of revisions.

Video Disc Mastering

Mastering the actual video disc still seems seat-of-the-pants (Steps 33-40), but gradually manufacturers are beginning to explain how producers can cooperate to get a quality product. The turnaround time for the optical video disc is still highly variable. Check discs are often returned within two weeks. The first mastering of most interactive video discs has errors that have gotten through the review, but remastering takes much less time. The manufacturers are expanding their capacity to press discs and to offer data preparation services so these delays will be reduced in the future.

FUTURE PRODUCTION

Microcomputers and video discs can now access up to 54,000 television frames with stereo sound. The computer terminal can superimpose high resolution graphics and text over the quality video picture. This combination results in a program that can be updated with data base content. In fact, a new medium has been created for electronic publishing, with its own rules for authoring, and with unique communication abilities.

In the future, we can expect much greater capacity for a larger variety of digital data, which will coordinate increasingly complex interfaces (connections with different types of equipment). During the next decade, touch sensitive displays, dynamic audio and gesture responsive programs (i.e., programs that respond to viewer movement, such as a nod) will require even higher sophistication in data preparation and digital techniques. Program support data will help structure more viewer-oriented equipment links that may achieve a higher logic or artificial intelligence. Naturally, software this complex will require unique approaches to simulate and evaluate programs.

Appendix 6A: Interactive Video Production Process

KEY

Note: Steps requiring staff personnel are marked as follows:

Production Assistant	**PA**	**Content Expert**	**CE**
Director of Photography	**DP**	**Producer**	**PR**
Character Generator Operator	**CG**	**Writer**	**WR**
Editor	**ED**	**Instructional Designer**	**ID**
Video Tape Operator	**VT**	**Computer Programmer**	**CP**
		Director	**DR**

PRODUCTION PROCESS

Staff Responsibility	Steps
1. Project Manager and Staff (CE, PR, WR, ID, CP, DR)	Hold preproduction planning conferences to set the production format, review the script, assign text redevelopment, models, graphics and video motion shots in a final visualization, review creative options, outline program map (preliminary flow chart for data preparation.)

Appendix 6A: (cont.)

2. DR, PA, CG	Produce text frames on the video character generator. Confirm the text content with the script.
3. CE, ID, DR	Review the text frames on video cassette for position, layout, and continuity.
4. PA	Proofread the text in Tell Frames and Videobanks.
5. CG	Correct the text in Tell Frames and Videobanks.
6. CE, ID	Approve the text, layout, and continuity of Tell Frames and Videobanks. Sign off on shooting script.
7. DP, PA, VT	Prepare art, slides, video tapes, films, or data recordings for still frame recording. Transfer all media to video tape and make time coded dubs.
8. CE, ID	Screen all rough cut materials. Suggest additions and changes in media.
9. DP, PA, ED, VT, CP, DR	Review all selected and modified materials for creativity, composition, interactivity, and continuity.
10. CE, ID	Sign off on all media materials and final shooting script.
11. DR, DP, PA, VT, crew and talent	Shoot the motion sequences following the shooting script exactly, and documenting shot record closely.
12. CE, ID, DR, DP, WR, ED	Get time coded dubs of motion recordings. Review quality and the need for reshooting, rewriting, or redesigning. Program simulation of segments.
13. CE, ID	Early formative evaluation of content with test sample, if there are problems.

Appendix 6A: (cont.)

14. ID, WR, DR	Revise and confirm the audio script for the narrator.
15. Narrator or talent	Record audio track.
16. DR, ED, PA	Edit audio track.
17. CE, ID	Review the edited audio track and make suggestions for improvements.
18. Narrator, ED	Rerecord segments of the audio track and add other effects into final track.
19. CE, ID, DR, WR	Sign off on the final audio track for the program from all of the content and creative perspectives.
20. ID, DR, CP	Change the preliminary flow chart to final form of the motion segments and still frames.
21. ID, CP	Write the computer program and debug the final version of the learning strategy.
ASSEMBLY AND PREMASTERING PROCESS	
22. DR, PA, DP, ED, VT	Color confirmation of all forms of media before the final assembly, and color correction if required. Confirm edit file and notes to eliminate extra time in assembly.
23. DR, PA, DP, ED, VT	Mix the text layouts and the picture media, and add the special effects to motion segments and still frames.
24. DR, PA, ED, VT	Make the master basic tape with time code, test signals, and cut in audio track.
25. DR, PA, DP, ED, VT	The final assembly can take several days of editing, but the edit list must be used to confirm the specific location of each piece of

Appendix 6A: (cont.)

	content. The PA will keep detailed logs that confirm the entry of each piece, and any changes that have to be made. The PR and the CP are on call or present during this process as well in case any changes have to be made.
26. VT	Make several dubs for the Project Manager and staff including CE, ID, PR, DR, PA, CP.
27. CE, ID, PR, DR	Meet with comments about changes or improvements in the edited assembly program.
28. DR, ID, CP	Send the master interactive tape to the manufacturer for a check disc in the video disc process. In video tape test the program with a preliminary sample of users according to the formative evaluation plan.
29. DR, ID, CP	Conduct a formative evaluation of the check disc with a valid sample of users, and document results in a report.
30. PM, CE, ID, PR, DR	Meet and review the formative evaluation results.
31. DR, PA, ED, VT	Correct the master video tape.
32. PM, CE, ID, DR, CP	Sign off on the master video tape for mass publication.
33. DR	Send the final master video tape to the mastering in volume.
34. ID, CP, PA	While the disc is being mastered use the check disc to confirm the preliminary computer program, debug it, and write the final program.
35. CP, PA	When the published discs return from the mastering plant, confirm data and dialogue. Change any computer code that is incorrect.

Appendix 6A: (cont.)

36. ID, PR, WR, DR, CP, PA	Review the total program with the entire creative staff before showing it to the CE.
37. CE, ID and Applied Researcher	Conduct the summative evaluation with a valid test sample selected by the ID or an independent Applied Researcher, and report the test data in a clear format. The CE will likely want to be involved with the learner verification process.
38. ID, CP	Revise computer code if required, and test each branch.
39. CE, ID, AR	Retest the revised program according to the summative evaluation plan.
40. Project Manager and staff (CE, ID, PR, DR, PA, CP)	Final review and sign off on the completed publication. Review the results of the summative evaluation process, and the production process.

7

Planning, Implementing and Budgeting Interactive Video Programs

by Steve and Beth Floyd

Interactive video is a new and complex technology, requiring considerable investment and commitment on the part of management. Obtaining approval and funding for an interactive video program may very well be the most challenging phase of the entire project. Therefore, before you make an appointment in the executive suite to share your ideas with management you must prepare a presentation that clearly justifies investment in the project. This presentation should accomplish the following basic objectives:

1) Define the tangible benefits for management
2) Highlight the major implementation steps
3) Clarify the cost centers and explain the budget requirements.

With these three points in mind, we will examine the critical steps for planning and budgeting an interactive video project.

PLANNING: DEFINING THE PROJECT

Identifying Objectives

If you work for a profit-oriented business, the first step is to demonstrate how the project will enable management to achieve organizational objectives. The organization might be a corporation, a division, a department, a plant or a regional office. You need to focus the benefits on the areas for which management is held financially accountable. In other words, you should plan to demonstrate how *an investment in this project will lead directly to tangible results* such as reduced operating costs, improved productivity or increased revenues.

For example, a training program for service technicians might help the operations manager improve productivity. A program on handling complaints might help reduce the average time per service call by showing technicians how to diffuse a problem situation quickly, before taking care of the equipment. However, for either of these programs to have credibility with management, the presentation must define measurable indicators that the manager can expect to result from the project. Ideally, these results will match the same areas that top management uses to track how well a manager is doing. The data might be increased return on investment, reduced overhead, improved operating margins, reduced costs or simply helping the manager find a way to stay within budget. Regardless of what form the data take, the objective is still the same: to demonstrate tangible results in terms that relate specifically to management objectives.

Not all projects lead to an easily measurable return on the investment. In fact, on first examination many projects cannot be tied neatly to a specific profit or cost center; rather, they affect several areas and may have other "hidden" benefits, such as improved corporate image.

For example, suppose you wanted to introduce an interactive video system in the corporate lobby, to perform a number of functions normally delegated to a receptionist. When a visitor entered the lobby, he could walk over to a large video monitor which displayed a range of choices such as:

- a brief history of the company
- an overview of recent accomplishments
- an update on financial performance
- an executive presenting the company's position and answering questions on a proposed government regulation
- a list of department telephone numbers
- the latest quotation on the stock market

Here, the visiting business customer or prospective employee could query the system to locate a phone number or to learn more about the company.

This type of a simple interactive system has several benefits. First, it conveys a positive impression of a progressive company. Second, since the system can be programmed to answer frequently asked questions, it enables the receptionist to perform other duties in addition to greeting visitors. Third, downtime for orienting a new receptionist is reduced, since the interactive system will handle most of the inquiries. By introducing a very simple interactive system into its lobby, an organization can take advantage of a public relations opportunity while making a receptionist more productive. A similar system could also be placed in a cafeteria or employee lounge to answer questions about employee benefits or to explain company policies.

Identifying Needs

In order to justify this type of project to management, the need must be clearly defined. That is why it is so important to analyze an organization's need in terms of a process, as we discussed in Chapter 1. The process approach relates projects to specific functions and needs. It might be nice to have an interactive system in the lobby, but if it doesn't solve a problem or meet a need then chances are that projects that meet more immediate needs will be funded instead.

It is important to describe the project in terms of a manager's areas of accountability. It doesn't make sense to convince the personnel department that an interactive system is needed in the lobby if, in fact, the public relations department is responsible for its operation (unless personnel is going to help justify the project to the public relations staff). A concensus approach is obviously valuable but make sure that you're presenting the idea to the right decision maker! (This can be facilitated by discussing an idea with successful innovators within the organization. Ask them for input about how to justify a particular project and who to approach for funding or support.)

Analyzing the Project

The example of placing an interactive video system in a corporate lobby raises an important question: Why produce an interactive video program if a bulletin board or a brochure can do the same thing with less cost? The answer requires some analysis and possibly some research. Taking the example given, the main issues to consider would be:

- How are the functions handled now?
- Is there a need to improve the way they are handled?
- Would other less expensive alternatives accomplish the objective?
- In this particular example, would the receptionist really be free to do other work?
- Is there enough traffic in the lobby to justify the public relations investment?

In order to answer these questions you must analyze the alternatives, defining the problems, needs and costs involved with each. (Table 7.1 presents this type of analysis for the example given above.) In this way, you can ensure that the interactive video project really does fit the specific needs of the organization, and is not just a nice, but expensive, proposition. This type of discipline should be followed on every project to make sure that it is really needed, and to identify specific areas where interactive video can make a unique contribution.

IMPLEMENTING: OUTLINING THE STEPS

After determining how to justify the project in terms of measurable results, the next step is to outline the major phases for implementing the project. This step has several important benefits.

1) It clarifies exactly what you plan to do and how you plan to accomplish it.
2) It helps you uncover potential problem areas or hidden costs.
3) It enables you to assign times and costs to major steps.
4) The plan demonstrates to management that you have analyzed the project thoroughly and that you are prepared to implement it.

In order to help you plan, we have organized an interactive video project into ten distinct steps:

1) Front-end analysis
2) Instructional/evaluation strategy
3) Instructional flow chart
4) Visualizing and scripting
5) Production
6) Assembly and programming
7) Pilot evaluation
8) Revision
9) Duplication and distribution
10) Project evaluation

These steps were selected for several reasons. First, they provide specific points for management review and approval before going on to the next action. Second, they divide the project into manageable cost centers for better cost control. Finally, each step can be managed or contracted separately, so that different groups or individuals are responsible for different functions. This may not be desirable for some programs but for a large

Table 7.1: Project Analysis Worksheet: Corporate Lobby

Alternatives	Problems	Costs	Benefits
Receptionist	Needs training Already busy Nonproductive	$12,000/year	Personal contact Security
Brochure	People don't read them Expensive to print Providing adequate copies	$10,000: printing graphics design	Tangible Reference guide
Bulletin board	Unattractive Doesn't fit decor	$500: installation	Easy to update
Interactive video	Cost to develop and produce	$10,000: develop using existing linear material	Public relations One time cost

project, or one requiring a team effort, it might help to clarify operating responsibilities and discover potential problems before they undermine a program's effectiveness.

We will briefly review each major step to highlight some of the critical decisions you must make as you put together a program. Next, we will review the costs that you can expect for each step as you prepare a budget. (Actual design, production and evaluation are covered extensively in previous chapters.)

Front-end Analysis

The first step in designing *any* program is to clearly define the problem you are trying to resolve, and this is particularly important for interactive programming since it provides a strong framework for organizing the material. The problem defined might be to reduce turnover, accidents or downtime. It might be to increase sales or line efficiencies. The important point is that you are trying to resolve a problem which can be measured in terms of return on the organization's investment. If the program doesn't help resolve a problem or improve performance, it is difficult to justify. With the problem clearly defined, the next part of the front-end analysis, the instructional design process, can proceed.

The instructional design process includes defining the training objectives in terms of *what* the trainee will *do* differently as a result of the session. Using a technical training program for service technicians as an example, an objective might be: "Following this segment, participants will list the procedures for calibrating test equipment." In this objective the "what" is calibrating and the "do" is listing the procedures.

The next part of this process is to analyze the trainees' entry level skills, knowledge and attitudes. This requires getting to know the trainees, how they perceive their jobs, what their expectations are, as well as how they actually perform their jobs. Next, conduct a task analysis to break the job into its major component parts. Organize the training around the critical skills or information the trainee needs in order to perform the job. You should not proceed until this predesign research is completed; trying to design a program before this preliminary background is complete will lead to false starts and create problems throughout the development process.

Instructional/Evaluation Strategy

The next step is to determine an instructional strategy. When you conceptualize an instructional strategy you are defining the general direction of the lesson. For example, do you want to make the program a highly interactive simulation with a multitude of branches or do you want a straightforward, linear presentation of content, controlled by the viewer's response to pretests and post-tests?

With the general direction and level of interaction in mind, organize the material into major groups; this will form your segments for branching. This step is directly dependent on the relationship between the learner analysis and the task analysis conducted earlier.

Remember the sales training example we referred to in Chapter 1? We looked at three levels of interaction: programmed instruction, programmed simulation and exploratory simulation. Using front-end analysis will help you determine how to match the type of learning and level of interaction with the trainee. For example, if product knowledge is the main problem, then the programmed instruction approach, which focuses on the learner's specific need for information, is indicated, rather than simulation, which presents situations requiring judgment based on product knowledge.

Once the instructional strategy is developed, you must develop an evaluation strategy, deciding how and when to evaluate a trainee's proficiency. For example, if you want to use a pretest, do you need one pretest for the entire program or a series of pretests before each segment? How do you want to structure the tests? Do you want to simulate a skill, such as landing a plane, or would you prefer to measure the learner's comprehension of content? Once the evaluation method is decided on, the critical decision points are placed in the program, based on the task analysis and the learner analysis.

After answering these general questions on direction, you have to define an acceptable level of demonstrated competence: what response or level of response enables the trainee to proceed to the next sequence, and what action requires branching the trainee through additional material. These are difficult questions, best answered by a careful review of the front-end analysis and evaluation strategy.

Next, review the content with a subject matter expert to identify common problems or potential trouble spots, and balance the expert's opinion with input from the trainees and their supervisors. Then determine where trainees will need additional remedial support or whether they simply need to recycle through the segment again for clarification. (This is especially important in technical training areas, so try to approach the material and the structure in the same way the trainee would encounter it.)

Finally, outline the program, identifying the major content segments, decision points, branches, simulated action, remedial loops, tests, hints and any other elements you plan to include in the program.

Flow Chart

The instructional flow chart serves as the road map for the program. Think of the flow chart as simply a diagram or blueprint of the instructional strategy. Worry about budget later—if you set too many limitations you may miss an opportunity to structure the material more effectively and more simply. (Several basic flow charts are illustrated in Chapter 1; more detailed flow charts are presented in Chapter 4.)

Visualizing and Scripting

When you have completed diagramming the instructional flow chart, confirmed the logic behind the decision points and reviewed the alternative branches, you are ready to begin visualizing and writing the script. As mentioned previously, writing an interactive video script uses all of the techniques you would normally apply to a traditional linear script. Follow the same process you use now (if it is successful) with additional emphasis on *visual continuity.* Always storyboard all of the possible paths, and try to see the program sequences as the trainee will see them. Check each branch to make sure that you have included all graphics, questions, motion sequence instructions, prompts, procedures, etc., that will appear on the screen in the storyboard. This also helps to organize the material.

Actively enlist the content expert's support at this point, to help with the branching strategy and to test the logic behind the structure of the program. Walk the content expert through all possible branches to discover where additional material is needed. Finally, match each visual in the program to the flow chart and correct any errors. This review with content experts will help reduce time-consuming patchwork in post-production.

Production

When producing an interactive video program pay extra attention to visual continuity. You may want to use symbols or color codes to help guide the viewer through the lesson. Do not compromise on quality. The amount of time and money invested in preproduction and design may tempt you to try and cut corners during the production to stay within budget, but resist this pressure or you may jeopardize your earlier efforts.

Assembly and Programming

When the graphics are inserted and the production segments are edited, the program is ready for assembly. This procedure will vary depending on the hardware system and distribution format you plan to use. Accuracy and organization are the keys to this step. (See Chapter 6.)

If you are going to use video disc, you should try a prototype test run with video tape before sending the assembled tape master to the disc mastering facility. This test run will help eliminate branching errors and serve as a check on the logic. If you are going to distribute on video tape, do a final paper-storyboard test to ensure that you have not missed a branch or constructed any deadend branches. Finally, the program is assembled according to the authoring system, or, if you plan to distribute on video disc, it is turned over to the disc manufacturer for mastering.

Pilot Evaluation

After the tape or disc has been programmed a pilot field test is set up before duplicating the program for release. During this evaluation, test all possible branching alternatives to make sure that you have eliminated any errors. Complex branching strategies require more thorough testing since there are more variables that can go wrong. This can be a tedious process but it is critical that any potential errors be corrected before the program is released.

"Debugging" (correcting errors) should involve three separate groups: the program designer, content expert and typical user or trainee. All three will approach the program from a different perspective.

The designer looks at structure, continuity and how the content fits together. The content expert needs to look at the overall structure, but also to review carefully any content discrepancies. Finally, the trainee or user will see how well the program works for him or her. Is it too easy? Does the structure make sense or was the program confusing? Were any sequences weak? The answers to these types of questions, combined with pretest and post-test data, will indicate the strengths and weaknesses of the program, and will enable you to measure the results of the pilot test.

Revision

The revision is a final clean-up to make sure that the program is workable and accurate. Input from the designer, content expert and users will help you decide how to revise the program to make sure it is on track. You may not need any revision; you may simply need to correct a few program errors; or you may need to conduct major surgery.

Duplication and Distribution

After completing revisions and checking them, you are ready for duplication and distribution. If you are using video discs, the duplication process will take approximately eight weeks to complete since duplication is handled at a central location for each manufacturer. If you send video tapes to an outside duplicator, it will generally take two to five days for turnaround.

Project Evaluation

After the program has been distributed to the field, compile a project evaluation to determine how effectively you solved the original problem defined in the front-end analysis. Keep in mind that evaluation is a continuous process of checks to review progress throughout the development of the program. Evaluation must be an integral part of the entire design in order to be effective. Feedback is often delayed until the project has already been implemented. This is counterproductive, since at that point it is generally too late or too expensive to revise the approach.

Currently, the 10 steps reviewed above are generally performed by the same person or by a small staff of two or three people. A few highly technical training projects for the military in weapons and aviation have separated the responsibilities for these functions into several teams, but this is the exception. One reason for examining these steps is to highlight the future need for a highly integrated management approach to interactive video programming. This approach requires that a different manager take responsibility for each different function. For example, the production manager is responsible for the production phase, while the evaluation team is responsible for evaluation and learning strategy. This does not mean that other managers are not involved or interested in the other phases. On the contrary, managers need to have overlapping responsibilities, and to have input into other areas. Similarly, design and evaluation teams can have overlapping responsibilities in design and the creative team can have joint responsibilities with production.

IMPLEMENTING: USING THE TEAM APPROACH

Let's look at how this integrated management approach might work. First, we start with a problem. A consulting team is assigned to review the problem and develop recommendations for action. Next, an instructional design team is brought in to conduct an instructional analysis and develop an instructional strategy based on the consulting team's research and their own analysis. After obtaining approval, the design team develops a flow chart based upon the design strategy. A creative team is called in to work on the flow chart and translate the design into a script and storyboard. The responsibilities overlap, yet each team brings a highly specialized expertise to focus on a particular phase.

The production team now enters the picture to shoot and put together video segments with action and graphics. The video segments, flow chart and design are handed over to the programming team, which assembles the interactive program according to the original design plan. Next, the instructional designers and content experts review the prototype program with an evaluation team. They conduct a field test with typical users and review the results together. If revisions are required they return the program to either the production or programming team for the necessary changes. After these changes are incorporated into the program, it might be reviewed again by several teams before it is duplicated. Once duplicated, the tape or disc is reviewed by a distribution team, which checks for quality and then releases the product to the field.

Finally, an evaluation team collects data on the project results to determine if it achieved its goals: solving the original problem. This is a very sensitive area involving both politics and egos. As a result, the make-up of the evaluation team will determine not only *what* is measured but *how* it is measured. Further, the interpretations of the results may be open to some debate, since it is difficult to set objective measurable standards. (This example is not necessarily typical of a single program approach, where one or two people might perform all of these skills. The actual number of steps or people involved will vary among projects.)

This type of management approach streamlines the interactive video process, while placing the overall project manager in a position calling for less technical knowledge but greater management and operations background. It is not as important to implement this kind of management structure to develop interactive programs as it is to keep in mind the collaborative team concept of this type of project.

BUDGETING: ASSIGNING THE COSTS

Adequate budgeting for an interactive video program is as essential to its success as proper planning and production. Developing interactive programs requires careful budgeting of each step in the process. Furthermore, since this is a new technology, some delays and problems have to be expected, and budgeted for—a concept to which we return later.

To illustrate the framework of an interactive video budget, we will present one, using the example of a new equipment training program for service technicians. Each of the 10 steps reviewed above are incorporated, and three approaches are compared: a traditional linear video program, a programmed instruction video program with some basic interaction and a sophisticated interactive program with programmed simulation. All the budget costs associated with each line item are assumed to be contracted outside the organization, to eliminate hidden costs and to facilitate comparison of the three approaches.

The linear program is divided into four 30-minute classroom lessons: operating controls, set-up and calibration, maintenance procedures and troubleshooting. Each lesson uses a ten minute video module to present the teaching points and demonstrate the skills. A classroom instructor follows the video module with exercises and discussions. The linear video program is supplemented with a leader's guide for the instructor, and workbooks with exercises for the participants. The linear project includes: 40 minutes of video tape, a 30-page leader's guide and an eight-page workbook. All video is to be mastered and edited on 1-inch "C" video tape using three-tube broadcast cameras, a professional crew and computer editing. As stated above, all design, production and duplication are contracted outside the organization.

There are almost an unlimited number of ways to design the two interactive video programs. For our purposes, the programmed instruction project begins with a brief pretest for the entire four modules. The participants are then branched to the appropriate module. Following each module the participants work through a series of questions. When participants demonstrate an acceptable level of competence they are routed to the next module. If they score below this level they are recycled through the module. This basic approach can be adapted with few changes from the linear program.

For the programmed simulation example, the participants learn by doing. Instead of progressing all the way through a module before encountering questions, each module has questions located at critical decision points throughout a sequence. The program then simulates the action selected by a trainee. This type of approach requires shooting and editing additional video footage. In this example we will extend the original linear video footage of 40 minutes to 60 minutes to take into account all of the additional branches and simulations.

The Comparison

With these descriptions and assumptions in mind we used the 10-step plan to relate costs to activities. Table 7.2 lists some average costs which could be associated with this project. The costs presented assume that each of these services is supplied without major complications or delays.

Notice that the linear program does not include a line for the flow chart or the assembly/programming step. However, a print and design category (for writing and printing the leader's guide and workbooks) has been added for the linear program. The interactive programs do not include this cost since the items are not necessary. In general, linear programs tend to neglect pilot test and project evaluations; however, we have included these steps for all three examples since they should not be limited to interactive projects.

It might surprise you to note that the total cost for the linear project (including the leader's guide and workbooks) is actually greater than for the basic programmed instruction example. The reason is simply that research, analysis and design take approximately the same amount of time, whether you develop exercises for a CRT or a workbook. Although programming and assembly add some costs to interactive programs, these are offset by preparing and printing the leader's guide and workbooks. You might think of assembly as typesetting and layout; thus the decision is whether to use paper, tape or disc.

Two important factors should be kept in mind as you review these comparisons. First, hardware costs for setting up the interactive systems were not computed for the individual project costs. These costs can run anywhere from less than $1000 to $5000 per location, depending on the sophistication of the system. (See Chapter 3 for discussions on hardware.) The second point to remember is that the figures represent national averages based on our experience with outside contractors for these services. The figures may vary widely depending on how the services are contracted and managed, but the comparison ratio between the three examples should remain valid.

The greatest cost for interactive video programs—video production—could run much less depending on the level of production involved. Keep in mind that we used a $1000 per minute cost guideline. Owning the equipment, or shooting a simpler program, would greatly reduce these costs.

Table 7.2: Comparative Budget: One Program, Three Formats

Step	Linear Video	Programmed Instruction	Programmed Simulation
1. Front-end analysis	$ 2,500	$ 2,500	$ 2,500
2. Instructional strategy	1,500	2,500	5,000
3. Flowchart	—	1,000	2,000
4. Visualizing and scripting	5,000	5,000	7,500
5. Production	40,000	40,000	60,000
6. Assembly and programming	—	5,000	10,000
7. Pilot test	1,000	1,000	1,000
8. Revision	1,000	1,500	3,000
9. Duplication	3,000	3,000	6,000
10. Project evaluation	1,500	1,500	1,500
(Print design—linear video only)	10,000	—	—
Total	$65,500	$63,000	$98,500

(Keep in mind that line item costs may vary, but comparison ratio between formats should remain consistent. For example, if you can develop a linear video program for less, by cutting production costs, then you would have to assume lower production costs for the other two formats as well, in order to compare them.)

The project manager might also be able to assign some of these responsibilities to personnel within the organization at no measurable cost to the specific project. Or, a supplier might bid on several services, providing a lower price than when contracting individually for each service. The costs presented are simply guidelines to use for comparison, not specific budget cost figures.

The point to emphasize is that simple interactive programs do not necessarily cost more than comparative linear programs. The cost difference in the programmed simulation example is primarily a factor of the increased production cost; additional video footage and programming are necessary to structure the alternative branching paths. Remember that in programmed simulation there is a greater number of branches. Therefore there is more material to shoot, assemble, store and duplicate. These production costs begin to escalate rapidly as you work with more complex interactive programs.

You should also note that we arbitrarily set duplication costs at $3000 for the first two examples and $6000 for the third. These video tape figures were used since a disc would not add any unique capabilities to the basic interactive example. In addition you would encounter some floppy disc duplication costs for interactive programs if they used a personal computer. Either video tape or video disc could be used for the third example, but because of the amount of video material (60 minutes) both sides of the video discs would have to be used for the four modules, while one VHS or Beta cassette could be used to reduce duplicating costs.

It is important to review carefully the duplication alternatives for different formats before completing the design strategy. Otherwise, you may miss some creative opportunities for reducing costs. For example, the still frame capability of the video disc might enable you to compress motion sequences to a few frames by stepping through the critical points of an action. This technique could eliminate many of the video motion sequences, so that all four modules could be packaged on a single video disc. Considering this technique in the learning strategy could reduce both production and duplication costs significantly. Use the reference guide in Chapter 2 to make tape versus disc comparisons. By using some of the unique capabilities of the video disc, and the savings from mass disc duplication runs, you may actually reduce costs.

To return to the caveat mentioned earlier, any creative endeavor has a vast number of variables that need to be planned for and managed. Interactive video is certainly no exception. Because the process is new (and therefore the cost variables are to a certain extent unfamiliar) you should plan to encounter some unanticipated costs and delays as the project unfolds. Plan to budget some additional funds to cover the costs of "the learning curve.'' This discretionary fund takes the pain and frustration out of exploring alternative sequences and branches. By allowing yourself the flexibility to experiment, you will also find new ways to save time and money on future projects.

The Proposal

Now let's use the technician training example to look at one alternative for making a presentation to management. Figure 7.1 highlights five areas for a proposal:

- situation
- problem
- project objective
- steps
- cost

This table requires you to analyze the potential benefits and costs for different alternatives, in order to focus the data on the immediate problem and possible solutions. (Of course, one real alternative in every project is simply to do nothing about the current situation. This often occurs because the cost of solving the problem outweighs the benefit.)

Figure 7.1: Technician Training Proposal Summary

Situation

No consistent training exists for service technicians
Standards vary at each location
Productivity varies between locations
High turnover rates
High frequency of repeated service calls

Problem

Operating costs are increasing by 20%
Supervisors and experienced technicians are spending 20-30% of their time training new technicians
28% of service calls were repeats of earlier calls to correct service errors
Average number of calls/man varies by as much as 30%

Project Objective

Train service technicians so that they can demonstrate procedures for operating controls, set-up and calibration, maintenance procedures, and troubleshooting for new test equipment.

Steps **Costs**

(Refer to Budget, Table 7.2)

As you conduct a cost-benefit analysis for a particular project, try to think of each cost center as a process. This approach will help you apply the technology creatively to take advantage of the potential savings described (eliminating printing costs, freeing up instructors, etc.). For example, even if it costs your organization 20% more to develop an interactive video project, perhaps it can save the organization 10 times that amount by reducing other costs. One project might even save the company enough in operating costs, travel and downtime to justify investment in an entire network.

SUMMARY

It is necessary to analyze three essential issues for management before trying to justify an interactive project:

- tangible benefits
- plan of action
- budget

As you develop a plan and budget for an interactive video project, use these three areas to help focus the presentation on results.

Regardless of the medium or the sophistication of the project, if it is thoroughly planned and carefully managed toward making a return on the investment, you should be able to make a measurable contribution to the organization. In the final analysis that is what planning, implementing and budgeting are all about.

8

Future Directions: The Curtain Rises on Interactive Video

by David Hon

The future direction of interactive video is probably not linear: that is, that future will probably branch and branch again. Unlike a well-planned flow chart, however, some of these branches will grow very strong and others will wither. So in talking about the future one should not be concerned only with the variety of possible directions—there will always be plenty of those—but with the relative strength of the directions which have begun, and of those which may begin soon.

For that reason, it would make sense to categorize the kinds of directions that will be emerging, and the simplest way might be to classify them as hardware and software directions. However, the first thing we must observe about any future is that not only the facts, but the *classification* of those facts, will not adhere to traditional categories. One only need look at the career planning of high school students in the 1950s to see that one very strong categorization of future jobs was men's jobs and women's jobs. Today such a categorization is passé.

So, even as we have just come to accept hardware and software as categorizations of development in interactive video, the terms may nearly be outmoded. Any good programmer will tell you that the right programming can make a video tape or video disc player capable of sequences and patterns you never imagined. Is inherent hardware capability then to be described as hardware plus the right software management? On the other hand, say an organization, such as the American Heart Association, creates a part of its software program to rely on inputs from an electronic training manikin which, in turn, cannot be used except with that software.

Is the electronic manikin then not also software rather than hardware?

The boundaries blur. And the first order of business is not so much to find out where you are, as to find out where everyone else is, because that is what spells your survival.

Until now, people have made their boundaries between hardware and software—not applications. May I suggest then that we look at categorizing futures in interactive video in a slightly different way than is now being done: that we look at those developments which fall in the area of technology (hardware and software) and those which fall in the area of applications. The reason for this division is that we have seen new technologies, fraught with wonderful, fertile possibilities, sit unused for years. For example, the video tape cassette took 10 years to enter into common usage, though the applications were there all along. And we continue to see tasks that can be done more easily and cheaply with a technological medium, done by very old means.

The point is this: the *presence* of technology does not mean the *use* of that technology. And that is probably quite a modern phenomenon. In the past, much technology evolved to solve problems. People knew they had a problem seeing in the dark, and that the smell of kerosene lamps, or the danger of gas lamps, were problems of only a slightly lesser order. So when the electric light was invented, people knew immediately what to do with it. The technology had been focused on solving the problem.

Today, technology may come into existence before people really have figured out what to use it for. Nuclear power is an exceptionally good example. The video disc is another. Very often in this century, technology has been ahead of its users. And, full of potential, fraught with implication, there it remains—too expensive to use (solar cells) or too dangerous (nuclear power) or just not understood in conventional terms (the computer, video discs).

The problem to date has been that technologists were allowed to think that their new hardware (or software) developments automatically equalled new realities; however this is not so. As magical as a technology may seem when it is finally used well, the roads and bridges from technology to its useful applications do not appear by magic—they are difficult and often frustrating to build. And more than one company, set to grab the interactive "window" (that is, when public acceptance and available technology create irrepressible demand) went under in 1981, waiting by the telephone.

Let us consider, then, that a new technology in the future may be the application of technology. The connection between technology, which has nothing going for it but its potential, and real-world applications, which are not dependent on that technology, needs to be made by a new kind of engineer; one who builds bridges between potential and reality. Therefore, the two categorizations that may be the most appropriate for interactive video, for the near future, are technological advances and real-world applications.

TECHNOLOGICAL ADVANCES

Technological advances will occur rapidly in video tape players, video disc players, monitors, microcomputers and peripherals. And, at the core of these advances, the development of more miniaturization and more memory per penny per inch. This means that if we have designs on the future we can be sure of the following:

1) *Lower prices* for any microprocessor-based systems. Volume production will make prices drop, in the manner we saw occur with the pocket calculator and the digital watch.

2) *Smaller systems,* such as pocket video tape players.

3) *Increased capability, durability and quality,* for example, longer playing video discs, more durable tape players, very high resolution flat screen TV.

4) *Video disc system sharing* enabling several users to timeshare on a video disc—a capability unique to video disc systems. The solid state laser heads, timesharing on a video disc, will make possible algorithms of usage that cable television people will be interested in.

5) *Rapid replication and recordable video discs.* First of all, rapid and dependable replication seems to be key to both custom (small run) applications and mass market applications, so you can be sure there is progress already being made in the replication area. Recordability in a video disc is quite a different matter. It is possible that if we have recordable video discs too soon in the future, the main protection that a program would have from duplication will not be possible. This takes away the incentive of a video disc maker to have his material safe from piracy. So if recordability comes too soon, it may actually hamper the large expenditures that will be necessary for truly interactive programs on the disc.

It seems that because someone mentioned video discs ought to be recordable, like video tapes, the engineers are now dutifully inventing that technology. Gutenberg was never asked if he left blank pages in his Bible so you could write your own inspirations in. Somehow that would have been presumptuous. But today he would have been asked, "But does it record?"

6. *Still frame audio,* which theoretically could apply to both video disc *and* video tape, because video tape freeze-framing is improving by the year. What this technology will allow is exponential expansion of content on the media. At 30 frames per second, 10 minutes could contain many good slide shows. The word concatenated will become important, because otherwise the still frame audio will require the narrator to pause for a split second while the frames change; a planning nightmare.

7. *Interface devices,* such as keypads and keyboards. The way in which a user communicates thoughts to the interactive video system is an important matter, and yet this area doesn't show strong development . . . *yet.* However, there haven't been millions of users shouting their preferences . . . *yet.* But soon this rather neglected, half-hearted collection of keyboards and touchpads and light pens and touch screens and voice recognition devices (as well as some other peripherals) will become the hottest path of technological advance. Right now most users don't know they have any say so, because they've been taught to sit and watch linear video and be glad of *whatever* interactivity came their way.

Just as the games arcade people recognized that joy sticks, roller balls and other devices heightened player participation, so too will interactive video designers realize this, as soon as they begin to listen to users more. And hear some of the same quarters drop.

8. *Authoring languages* which allow broader programming creativity for microcomputer-linked systems. Because the technological whole is often increased by adding individual components to the system, the microcomputer is both its own frontier and the component which allows system frontiers for video. "Surrogate travel" may require computer control of four or more video disc players, for example, with player #2 covering right turns down side streets of disc #1 and player #3 snapping in right turns off those streets. Figure 8.1 illustrates this.

If we drove north on Webb to Elm then turned east on Elm, then turned south on Caribou, and then west on Cherry, we would have used players 1, 2, 3 and 4. The computer switching would have been done at 1B, 3B and 3C and the scene in front of us would appear as it would if we were driving west on Cherry.

The Massachusetts Institute of Technology (MIT), the Department of Defense, and Videodisc Publishing, Inc., along with a number of others, have demonstrated the need for computer control of systems.

Authoring systems do exist for travel and archival applications, because the search activities are similar. However, search activities differ for other applications. For example, the American Heart Association has demon-

Figure 8.1 Surrogate Travel Grid

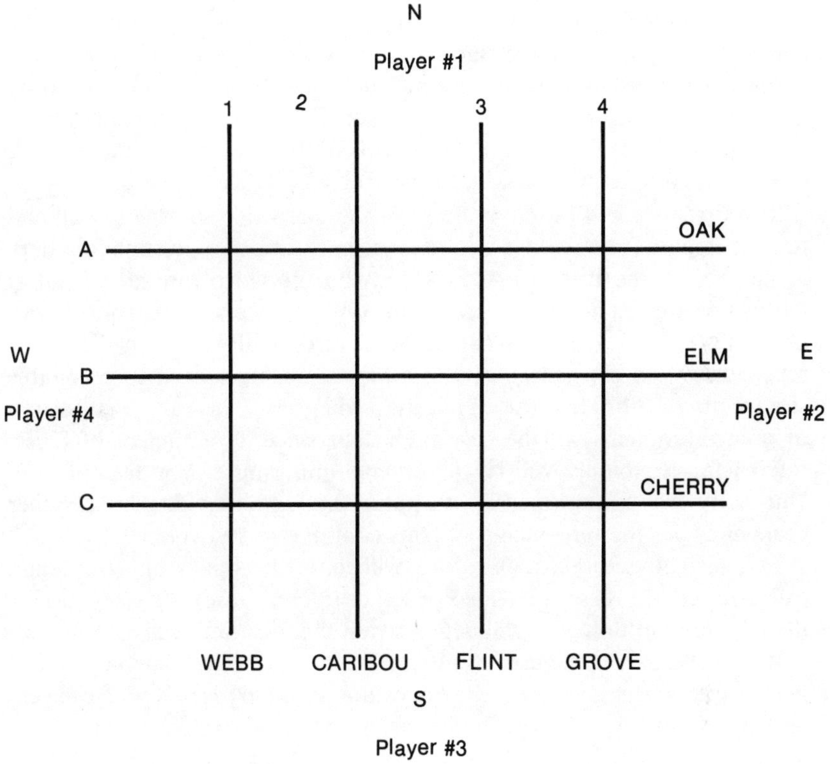

N

Player #1

1 2 3 4

A OAK

W B ELM E

Player #4 Player #2

C CHERRY

WEBB CARIBOU FLINT GROVE

S

Player #3

strated a video disc microcomputer with an electronic manikin as part of its teaching system. The system allows computer evaluation of hands-on performance and instantaneous coaching to be displayed on the video disc.

These uses and systems have just begun, but eventually there will be authoring languages with which the novice can develop applications for these complex systems. Most of the early authoring languages now in existence are slightly premature. They tend to be *deductive* before we've had enough *inductive* experience with the genre, and hence tend to hamstring users to simple formats and simple branching, such as multiple choice questions. On the other hand, current authoring languages are giving the novice computer capabilities never accessible before, so perhaps these authoring languages will evolve as applications proliferate.

The development of the very sophisticated authoring language is an important example of software's being a subset of technology, and not being equated with an actual application. An application is a problem solved, an opportunity exploited, a job done.

REAL-WORLD APPLICATIONS

When the deployment of technology finally becomes recognized as the new technology, we will see startling applications (developed by nontechnologists) which strikingly reduce time and money output, while increasing user interest and productivity. The current use of automated teller machines (ATMs) in banks is an example, but there will be many more which use video as well as computers.

The first uses will probably be *archival.* The video storage of still and moving images for reference purposes seems the most opportune first step. A number of the best efforts now seen on the video disc are aimed at harnessing the immensity of stored information we are confronted with onto 54,000 frames. But the patent search discs, realty discs, medical slide repositories and art gallery discs are now all being turned to profitable storage application, and this application will grow. The computer portion of archival systems will be very important since development of cross-referencing techniques will be of supreme importance. For example, the University of Washington in Seattle put a 20,000-photo video disc together years ago, but the referencing systems continue to be explored.

In the slightly more distant future, will come the applications we should call *dynamic.* In these, the video program (tape or disc) will truly engage the user, and in doing so will demonstrate the most interactive benefits.

It is in these dynamic uses of interactive video that we may have to call in the experts. But it will be more important to call in the real practising experts who can bring their vital dimension to the interactive medium:

- The teachers who can truly teach, whose students continue to excel whatever system they are in;
- The mothers and fathers who have motivated their children to work, and wonder and cope and sometimes, win;
- The entertainers who pack the seats night after night;
- The advertisers whose TV ads inform and motivate in 30 seconds;
- The managers whose people continually exceed expectations.

In other words, let's not look in the wrong places for the ways we contact the minds of users of the interactive medium. We have a lot of experts who have learned to empathize with, excite and get through to other human beings, and these are the experts we need to condition to the new media. The technology and the interactive design are secondary to the finger on the live pulse. If an interactive program cannot engage the user, and constantly stay in touch, all the other "expertise" is a tremendous waste.

So the dynamic applications of interactive video are the ones that will revolutionize the world of human communications. It seems as inevitable as the American move west that dynamic applications will venture farther and farther from the "authoritarian" linear one-way communication, to where the viewer and the medium will be equals. The participant will eventually be able to flip almost at will through a program.

Of course the book represents such a case already. The dictionary is archival and the mystery novel is dynamic, and you can legally open either to any page. But with interactive video there will be several more dimensions. First of all, there will be several levels of interactivity.

Level One Interactivity—Directed/Response

Currently most interactive video is being done in this level. *Directed* interactivity would mean activating by user choice, such as in point of purchase displays or video tapes that explain various personnel policies or benefits as the user chooses. *Responsive* interactivity occurs when the user is asked to select, recall or in some way perform for the program, material he or she has been exposed to. Multiple choice testing is of course in this level, but many more sophisticated means of evaluation will be also, before users demand to go to a second level.

Level Two Interactivity—Exploratory

As one might peruse a newspaper, whole subject areas can be "explored" on video disc or video tape and not just on a user-selecting segment basis. Already skilled disc users achieve humorous effects such as flipping divers out of the water and backing polevaulters off for another try. Once we decide to let users explore a subject, they tend to learn and experience in more personal ways. It is quite possible to allow maximum exploration and practice in a learning environment and still insist on standards. One might imagine a grammar disc, for example, which allowed a student to explore 15 different ways of saying the same thing, but held the student accountable for knowing a noun from a verb. Surrogate travel, mentioned before, could be a familiarization tool for new ambulance drivers.

Level Three Interactivity—Creativity

It is not unlikely that the most sophisticated computer chess programs might be so perceptive as to say to the player: "Alvin, that's the most brilliant chess move I've ever seen. I'll be mated in three moves and there's nothing I can do to stop it." Wouldn't that make Alvin feel superb? Or if one used visuals on video to construct a sequence which showed a command of the material, such as selecting various architectural forms used to support a bridge, level three interactivity would be flexible enough to allow a design the programmer had *not yet thought of.* Moreover, the computer could evaluate and compliment the user if the sequence or construction was both new and workable.

It may be a while before we see this third level, but perhaps not so long after all. Possibly the most interesting work now being done in both computers and interactive video is that which blends learning and entertainment. Not just learning which is animated and fun, and not just entertainment which skillfully carries messages, but games where you must learn to win, and wherein the attempt to become better at the game involves some necessary changes in perception.

One of the early computer games, Star Trek, did this exceptionally well. The player had to know the points of a compass cold to navigate and fire rockets at evil Klingons and to avoid their pursuit. There was no way to avoid it. In order to get better at the game, the player's navigational ability had to improve. The stimulus to learning was to outwit the computer by first learning navigation, and then by *creatively* outmaneuvering its programmed (and very well programmed) attempts to do you in.

It may be, then, that the users who learn to outsmart the intent of the interactive program (video, computer, etc.) are indeed learning the most about the subject. And perhaps our efforts should be toward rewarding this sort of creative behavior, playful though it seems, rather than penalizing it.

THEMES OF DYNAMIC APPLICATIONS

The creativity in future design will, at its best, give rise to user creativity. Along the way, a key factor will be the attitude of designers toward users. Themes that we choose will be engaging, and may allow the user to pre-experience a real-life situation. The beginnings of drama and the beginnings of education may well have been the same, perhaps when the older hunters "staged" a hunt before the campfire, involving the younger hunters so that they would be more ready for the actual hunt. The early Greek theatre evolved from this, but so, one suspects, did military war games. That both have lasted so long is testament to the power of pre-experiencing as a learning tool.

The danger here is that interactive video done extremely well could become a substitute for real life, in a way that combines the worst of the TV and video game addict of today. But let us put that in a category of dangers which will be social issues of the future. It will be difficult to produce interactive video that well for some time to come.

A future theme may be between one screen (with mixed text, video and computer graphics) or two screens (separating computer from video and allowing crossover reinforcement and counterpoint). This issue will probably emerge as something of true consequence in right brain/left brain literature, and will be the subject of numerous testing "showdowns" and doctoral theses.

The application of technology will turn out to be a much more thrilling and productive area than the technological development itself, and will obviously spark more advances in the technology. Unlike most analogies which do not characterize the technologist as a creative person, in this next future we might say the technologist is the playwright and the applications people are the director and cast. The director and cast are always looking around for a good play, but meanwhile they will produce the play at hand. And they'll be better off, of course, when the great play is written. But conversely, however great the play, it must have its applications people, the actors. Playwrights have rarely been so pompous as to assume that any actor could do as much justice to the invention as another. The great play —or the great invention—lives or dies by the craft of the players.

Those overly concerned with the implications of each piece of technology should remember: most people go to see the actors perform, not to hear the lines of the play being spoken aloud. That is what will make the future of interactive video bright: the technology *plus* the performance of those of us who will use it in the arena.

Appendix: Case Studies in Interactive Video

AMERICAN HEART ASSOCIATION: CPR TRAINING

by David Hon

Statistics reveal that every other person in the United States dies of a heart or blood vessel disease. Of those, 350,000 people who die each year of heart attacks do not make it to emergency care. Often scant minutes, even seconds, stand grimly between life departing and help arriving. In many cases cardiopulmonary resuscitation (CPR) is the only way to sustain a victim's life in that vital time when the heart stops and emergency care is still on its way.

Through the teaching of CPR, thousands of lives have been saved, many by nonmedical rescuers—lay people on the spot who knew what to do to keep a victim's heart and lungs going until help arrived. But a true outreach of CPR training classes to a substantial portion of the population could tax even the largest and wealthiest of organizations, let alone community-based organizations such as now teach CPR as regularly as they can.

Use of Interactive Video

The American Heart Association (Dallas, TX), which pioneered CPR and set the medical standards for all emergency cardiac care, has now found a way to achieve that massive outreach of CPR using a video disc within a learning system. The system allows organizations to increase programs without increasing their number of live instructors, while still maintaining absolute standards and instructions.

The truly new feature of the system is that hands-on performance is now taught by the "victim"—a manikin which is normally used in instructing live classes. It is wired with an array of sensors to give precise feedback to an Apple II computer. From the evaluation of performance, a proper "coaching" response is elicited immediately as the "victim" responds, causing performance to become more and more precise.

Design and Program Features

Other design features of the system include light pen control of a variety of choices from the computer screen; "talk along" instruction by random access audio to accompany freeze-frames; fill-in blanks by use of "word templates;" independence from the computer keyboard; a vocabulary check with a "word bank" of a few visualized definitions; and an end-user target price in the neighborhood of $5000. These features are discussed in more detail, below.

The light pen control on the computer monitor totally frees the student from the computer keyboard (which is, in fact, hidden). All interaction takes place through the computer monitor screen. A constant source of choices is offered on a nine point "choice menu." The student may call up the menu instantaneously to review (frame by frame if necessary); to challenge sections (or the whole course); to opt for practice in the more difficult areas; to ask for immediate definitions through the vocabulary check, or to explore material which treats lesson material in reference fashion; to take a break and return (a month later if necessary); or to resume the program at the point at which it stopped.

The word template system allows the recall method of evaluation to be used for key concept words. In this system, the light pen enters letters from an on-screen alphabet and the computer evaluates only key combinations. For instance, for the word "pulse" to be correct, when all the blanks are filled in the computer scans the answer to see if the letters PLS are there in that order. If they are, the answer is correct.

In addition, a few high-resolution graphics can be used for the student to indicate locations on the body. At this stage high-resolution graphics, as well as extensive text, may be limitations on the 48K computer because both modes require heavy memory usage. The ideal method is to have the microcomputer act as a "manager" of the mosaic course structure, and the 48K Apple II is quite capable in this regard.

Approximately 200 frames are stated for the vocabulary check. Most of the definitions are visualized and constructed in levels of complexity. The first definition is crisp and simple, allowing the student to understand quickly and return to the program. If the student chooses, however, he may "explore" most concepts in more detail with a touch of the light pen. This is part of the mosaic concept of learning. Recognizing that certain objectives must be achieved, this system also tailors the course to vagaries in each student's learning method and caters to deeper interests in certain topics as they arise. Also, this method allows levels of depth in this medical program for people who want more detailed medical definitions. In this way, several levels of learning can be satisfied with one nucleus program (something to think of when amortising front end expenses.)

The low target price is necessary to the broad outreach of CPR training which the American Heart Association would like to encourage. The figure would present few problems to institutions which would like to expand ongoing CPR programs but cannot justify the manpower or logistical resources. The system can be used for training from one to four people (individually evaluated) without an instructor, or as a rapid recertification tool for those who have taken a CPR course before. A conservative figure on the number of people who could be recertified in a year (if they had a course six months before) is 1700, if the machine were in use two hours a day.

There is no absolute time for completion of a mosaic course. The time to take the course may vary from one to five hours depending on prior medical knowledge. We estimate that the average for one person taking the total course might be about two hours; for four people, about four hours; for recertification after six months, about 30 minutes; and for a recertification after one year, about one hour.

BANK OF AMERICA: PILOT TRAINING PROGRAM

by Nicholas V. Iuppa

Bank of America has very diverse video programming—necessary to serve its 80,000 employees in 1100 branches. Each branch in California (and each floor of the administrative buildings) is equipped with a ¾-inch video cassette player.

In 1981 our in-house production facility turned out 30 programs for distribution throughout California, as well as another 35 simple "live-on-tape" programs for limited use. We have been very successful in our efforts and have received numerous awards. Yet we do experience a great deal of frustration. The frustration exists because, frankly, when it comes to training, *television can't each anybody how to do anything.*

Since 80% of our programming is oriented to training, that presents a real problem. Fortunately, with the development of interactive video, television will not only teach but may, in fact, become *the* most important teaching medium in the world.

To understand exactly how and why interactive video can revolutionize learning it is necessary to understand what it takes to teach somebody how to do something.

First, you have to be able to show your students what it is you want them to do: you have to *demonstrate* the skill. Television has always been able to show people how to do things; it's a tremendous demonstration tool. But demonstration is not really enough. To really learn, students need a chance to practice the skill they have just seen demonstrated. Until now, television has not been able to let you practice anything.

One of the greatest advantages of interactive video is that it allows student involvement. It has the ability to let people practice many different kinds of skills, especially cognitive skills like discrimination, generalization and sequencing.

Interactive video also provides the final ingredient for real learning: it allows you to test whether or not learners can perform the skill you have just taught them. It asks questions, tabulates scores and gives the results. This practical, easy-to-use hardware that can demonstrate and allow practice and testing never really existed before.

Use of Interactive Video

At Bank of America we looked forward to the emergence of this new technology. After a good deal of consultation with bank administrators, we decided to test interactive video by producing an original interactive video disc program on debits and credits. The program taught how to decide whether an item presented at the teller window was a debit or a credit. (This is one of the first things tellers must learn, yet many have great difficulty doing so.)

We expanded the subject by adding units on two other basic teller skills: how to balance your work and how to use the bank's stamps. This program has been completed and is currently being piloted in our teller schools and in some of our larger offices.

Design and Program Features

Let's take a close look at our first interactive video program. It begins with a standard explanation of the interactive system itself, and is followed by three content modules and a final exam. Debits and credits is a prerequisite to the other subjects. After demonstrating knowledge of debits and credits, the learner can choose the next subject to study.

The general structure of each individual lesson is: demonstration— exercise—demonstration—exercise—drill—review—test. There are as many demonstrations and exercises as the content dictates.

We used three major kinds of learning exercises. In the first the answer is fairly straightforward, so regardless of the choice made, the feedback is the same. The questions are not repeated, because the situation is too clear-cut; feedback explains all possible answers.

Our second kind of exercise is more sophisticated. If a student gives a wrong answer, the reason why the answer is wrong is explained, and the question is then repeated. Separate explanations are given until the student gets the right answer. (This kind of exercise is called a remedial loop.)

In the most complex exercise, the right answer takes you further into the program. A wrong answer, but one which is not too unreasonable, gets a simple remedial loop. An answer which is completely wrong gets a whole new remedial question. A student who misses that question is given the option of going back to the start of the lesson, since he or she probably didn't understand the basic concept to begin with.

After the lesson is completed, the student takes a short test. If he passes the test, he can proceed to another unit, to the final exam, or he can review the entire program by jumping to the review sections of every lesson before taking the final exam.

If he fails any test, he must repeat the appropriate lesson, either by seeing the review again, by going through the whole lesson or by seeing a special version of the lesson that skips the exercises and presents just the demonstration.

The final exam tries to present the fullest possible simulation of the actual skills being taught. It looks at check cashing transactions from the teller's point of view, and combines several decisions into each test item.

Evaluation

This interactive video disc program has been a very successful first effort, and is currently being tested in six locations. Evaluations of the program have been conducted with the following results:

Positive student response to questionnaire (from all students participating)

- 100% said exercises helped the trainee learn
- 100% felt information was clearly presented
- 97% said rate of presentation was good
- 96% felt important concepts were taught
- 96% said material was well organized
- 90% found style of production appropriate
- 83% had no machine breakdowns
- 75% said enough interaction was used

Positive teacher comments (from interviews with instructors)
- Can't teach without it
- Cuts teaching time
- Students understand principles; don't just memorize answers.

Improved productivity (not yet quantified)
- 100% efficiency at these skills first day of job versus one week needed to gain efficiency without interactive video program
- Improved quality of service (reduced errors)

Production preferences of teachers (techniques favored)
- Prefer narrator reading questions
- Prefer complex/varied interactions versus simple "question/repeat" formula
- Prefer simple, movable key pad
- Prefer teller school's group participation followed by individual use for slow learners (This major finding changes anticipated need from 25 machines per school to one per school.)

Conclusions

Our great success has been with a skills program. People have learned how to *do* something far better than they ever could with standard video. But what about subjective (or affective) activities? What will happen when the behavior to be taught is far more difficult to define? We are tackling that problem right now in our second interactive video program, on customer relations.

Our plan for the near future is to continue to develop software for interactive programs, to place our test modules carefully and to expand our interactive video network slowly (we have not selected one machine or one system). We continue to test and to compare, and welcome any system or product that thinks it can do the job better.

COLLEGE OF SAN MATEO: ELECTRONICS TRAINING*

The College of San Mateo's Worksite Training Project is using interactive video to respond to the employment-training needs of California's prominent electronics industry. In order to train employed persons to assume increased technical responsibilities in electronics ("upgrade train-

*Material for this case study was contributed by Bob Whitney of Whitney Educational Services.

ing"), interactive video and computer-managed instruction are used for competency training in industry-endorsed skill areas.

Design and Program Features

The computer-managed delivery system allows for company implementation of self-paced, modularized, open-entry/exit, competency-based individualized training for selected workers. Self-paced video lesson material uses existing and newly created video tapes. The interactive instruction is delivered by the Computer Assisted Television Instruction (CATI) system developed by Whitney Educational Services of San Francisco, CA.

The CATI computer interface and supporting software integrate an Apple II microcomputer and a small video tape player. The trainee interacts with the computer-based lesson material and is presented with remedial segments of video tape according to his or her individual needs. The instruction includes workbook exercises, review and testing, adapted commercial and military video tape programs, interactive hands-on exercises and skill tapes that emphasize measurement and troubleshooting techniques.

Another unique aspect of upgrade training is that it is delivered in identical training stations that are placed at company sites ("worksite" training). These training stations contain computer and video tape equipment, electronic testing equipment and actual operating electronic equipment types for measurement, calibration and troubleshooting activities. The company-selected "upgrade" trainees schedule time on the training stations at their convenience.

Identical training stations are installed at the college to accommodate trainees and companies who may find it more convenient to work on-campus. The upgrade training courses are self-paced, allowing trainees to proceed through the instruction at their own speed. Academic credits are awarded for the upgrade training on the basis of one full (calendar) year of instruction, but the credits are earned as the trainees demonstrate competency with the content of the instruction.

DISCOVISION ASSOCIATES: INTERACTIVE MARKETING DISC*

More than 1100 European buyers were recently treated to an innovative combination of interactive video technology and the latest fashions. Interactive video discs were used to preview new fashion lines at the 1982 "World Buyer's Week" in New York. The disc enabled buyers to review a

*Material for this case study was contributed by James Zinn of Pioneer Video, Inc.

designer's product line before meeting a supplier.

DiscoVision Associates (DVA) was asked to produce the fashion disc to support World Buyer's Week, only three months before the show opened. With less than three months to shoot, edit, program and master the disc, the team was forced to create as the project unfolded.

Shooting began in late March, with 30 to 40 shots per outfit. In order to meet the tight deadline, four fashion photographers worked each day. Each photographer worked with three models who changed daily, doing eight designer's lines per day with three to four outfits for each designer. The daily logistics were complicated, with problems ranging from delivering the right line of clothes on time, to model availability, to processing slides overnight. The photographers' work was edited the day after the shoot, so that the best slides could be selected for the disc.

The slide sorting process alone required an entire day for each photographer. After the shooting was completed, the slides were organized into sequences. One slide carried the designer's name, followed by four slides showing different outfits, followed by two logo slides which read "Welcome To New York" and "Press In Your New Video Disc Code."

The collection of 1500 slides was converted into a 35mm continuous filmstrip, with one image per frame. Next the filmstrip was transferred to video tape and edited on a CMX system to correct any mistakes and remove splices. Then the material was sent to the DVA plant for disc mastering. Thirteen individual kiosks were set up at the event for the disc presentations. Each kiosk housed one Sony Trinitron and a DiscoVision industrial video disc player.

Program and Design Features

In conjunction with the World Buyer's Week Show, foreign buyers visited the Parsons School of Design in New York. There, DVA set up a computer terminal, which enabled the buyers to access the disc. Interacting with the terminal, buyers were asked questions, such as: What language did they wish to converse in and have on their print-out (the discs were programmed for English, French, German, Italian and Spanish)? What types of clothes did they want to see? What sizes? What price range?

The computer then processed this information to select a few designers and addresses to match each buyer's requirements. This information was generated on a coded print-out which the buyers could carry to a kiosk. At the kiosk the buyers entered the code for one of the designers, and could then view that designer's fashions on the screen. This procedure enabled the buyers to screen a fashion line briefly without visiting each showroom in person. Not only did this measure save time but it also gave buyers access to a broader selection of designers' clothing.

FIRST UNION BANK OF NORTH CAROLINA: BRANCH AUTOMATION TRAINING*

by Dick Handshaw

In the summer of 1981, First Union National Bank of North Carolina became the first major corporation to implement an interactive video program managed by a TRS-80 microcomputer. The project didn't begin as an experiment in interactive video; it began with a serious training problem. The solution to that training problem became a pioneering adventure in interactive video.

First Union operates approximately 200 branch offices throughout North Carolina. The banking industry is characterized by frequent changes in procedure and policy as a result of fluctuations in the economy and changes in government regulations. This climate of change, coupled with a large, geographically-dispersed audience, creates a challenge for any training effort.

First Union is undergoing a project that every major bank will deal with in this decade. The project is called "branch automation." In banking today, information is the key to customer sales and service. Computers, of course, are the key to delivering that information.

Branch automation, when completed, will change most of the operational procedures in First Union's branches. The first phase of the project was the distribution of a computer system called the Customer Information System (CIS), to provide all non-teller personnel with up-to-date information on customer accounts. Thus, employees had to learn to operate the new computer terminals and become familiar with the system, in order to find information quickly and accurately. Further, the installation of CIS would introduce the automation concept and set the tone for delivery of future automation packages.

In analyzing the learners and learning environment for our training program, we identified the following needs and selection criteria. The training vehicle should:

- Show clear demonstrations;
- Provide consistent training;
- Be self-paced to meet different learning rates;
- Assure student performance and mastery;
- Provide for future revisions;
- Provide for student practice;
- Maximize investment over long term;
- Build a positive attitude toward the automation project.

In considering classroom or one-on-one training, there were two options: to bring employees to headquarters or, to send trainers to hub cities. The one-on-one method would meet most of our needs for the long term. However, new employees or procedures would require additional workshops. Additional computer terminals for training purposes would be required, and a great deal of travel and lodging expense would be incurred. Cost was a prohibitive factor for this method.

Use of Interactive Video

Instructional video tapes were an accepted and successful means of delivering some training at First Union. With video tape we could provide clear demonstrations without the purchase of additional terminals. A video tape alone, however, could not give the learner a chance to practice new skills or get feedback on performance. Also, there was no way to assure any level of proficiency.

With the concern for feedback and student management in mind, we looked at Control Data Corp's Plato system, but Plato is dependent on Control Data's mainframe computer, making it time-consuming and expensive.

We then considered IBM's Interactive Instructional System (IIS). It would operate on the very same terminals that would be used for our production system, and we could supplement the text of the computer-assisted instruction (CAI) program with demonstrations on video tape. But there were problems with scheduling the course writing, and the location of our terminals. We also realized that the need for the learners to manage the various pieces of video tape for a demonstration of each different transaction would become a nightmare for them.

What we needed was a system to simulate the computer that ran the CIS program. The system also had to be able to show a demonstration of each transaction as it taught the course. We discovered that the keyboard on the Tandy Model III microcomputer was similar to our IBM keyboard, and that we could match the input and output closely enough for our training purposes. When Whitney Educational Systems agreed to build an interface for our TRS-80 microcomputers, our system was born.

Design and Program Features

Development of an interactive video program offered some unique challenges. I was an instructional developer and video producer, but knew virtually nothing about microcomputers. I had to rely on the technical planning staff of First Union's computer company to recommend the software to use with the TRS-80. Their choice was Tandy's PILOT + (later renamed MICRO PILOT).

PILOT + is an instructional software program that allows educators to write their own CAI programs without the need for data processing skills. PILOT + was not available at the time we needed to begin our development, so the technical planning section gave me a BASIC program written from PILOT + documentation. I was able to use this to imitate some of the functions of PILOT +, and within a few weeks had learned enough to use it for writing interactive video programs.

We began with a five-month time allotment in order to produce the completed prototype. The first two and a half months were spent developing the first two modules—about one-tenth the total content. I wrote the script for the first video demonstration and programmed the computer to simulate the transaction shown in the demonstration. A great deal of time was spent testing this first module, until it was easily understood by the test audience.

We developed a takeoff on the "Mission Impossible" television show, which we named "Mission Achievable." One of our objectives was to make people feel comfortable with the concept of automation, so the entire program took on a light and friendly tone. The computer was especially reassuring, using phrases such as: "That's not quite right, but that's okay, Ted. Try again." We did add a little human sarcasm from time to time, programming the computer to say, for example: "I can't believe you forgot how to terminate."

We built a personality into our system. As we noticed our test students interacting with this personality, we decided to name the system Merlin. (In the legend of King Arthur, Merlin, the sorcerer, knew the answers to all questions. He would never tell Arthur what to do, but gave hints and information to help Arthur solve the problems himself. This, essentially, was our basic design strategy.)

The strategy for "Mission Achievable" was developed in the testing sequence of the first training module. We spent time explaining the use of the training equipment and the convenience it had to offer, continually stressing the enjoyment and ease of operation. Next, we gave a video demonstration of the most simple transaction on the IBM terminal, followed by a simulation on the TRS-80. The first simulation guided the learner through each new step. Mistakes could be corrected by typing "HINT" which would give a text clue on the screen, or by typing "TV" to review only the needed portion of the demonstration.

During the development of the first module, I learned that the script and text for the computer needed to be written from the content outline as one continuous dialog. I also learned that production of the computer program and video tape was not a one-person job. With one-tenth of the project completed at the halfway point, we organized a project team.

Our production problems were solved, but we still faced a major hurdle just prior to our implementation date. Because Whitney had to design and build an interface to our specifications, we did not have a working prototype of it until July 5, 1981.*

Because the program was written as a single entity, it worked just as planned and tested, prior to the arrival of the interface. Telling the computer where to find the video segments was simple because of this preplanning, and we were able to implement a field test in four cities on July 13, 1982, just 7 months after we began the project.

Implementation

The bank had chosen 12 cities, with a target audience of 100 people, for implementation and testing of CIS. This made an ideal field test for "Mission Achievable."

Each Merlin unit consisted of a TRS-80 Model III microcomputer with 48K RAM; a Whitney interface with cables; a Panasonic 8170 VHS video player; a Panasonic 13-inch color receiver; a two-shelf cart for player and receiver, and a small Radio Shack table for the TRS-80; and a ½-inch VHS video tape recorder and two mini diskettes.

Even though we color-coded cables, and made the configuration as simple as possible, we learned that someone with knowledge of the equipment actually had to set up and test the units on location. And, although we had tested the equipment, we did have our share of equipment failures. We managed to keep enough units working to train our target audience during the three-week field test.

Evaluation

In order to establish a standard for measuring the success of our training program, we consulted with our Customer Information System department. (This department uses trained operators to access CIS and handle questions over the telephone.) Training in CIS was done on a one-to-one

---

*The interface uses a time code generated by the TRS-80, which is recorded on an audio channel of our ¾-inch master tape. Our interface does not just count control track pulses, but runs on an addressable time code similar in function to the SMPTE time code. This is a key factor to search accuracy and ease of producing multiple copies of the program.

basis, and required half of a productive employee's time for the two to four week training period required to bring a new operator up to production capability. A three-month training period was generally required to bring new operators up to peak efficiency.

We sent evaluators with each Merlin unit, and discovered that an average time of 3 hours and 10 minutes was spent with the training material. The training usually took place over a three-day period, with a total of 90 minutes practice on the IBM terminal at points throughout the program. Learners were evaluated on their mastery of each transaction on the IBM terminal. More than 80% of the test audience was rated GOOD, 17% was rated FAIR and 2% was rated POOR.

We originally scheduled learners to undergo two-hour training sessions on three consecutive days, but found that one and a half hour sessions were more productive. With only one demonstration of the training equipment, 95% of our test audience found Merlin easy or fairly easy to use. We also found that about half our test audience elected to repeat "TV" segments when they made mistakes, while the other half used only text "HINTS." Sixty-four percent of the learners required some assistance from the evaluators at some point in the program. We believe we can significantly reduce that percentage with some revisions to make the training program truly automated.

Conclusions

A significant benefit of "Mission Achievable" was the apparent cure it provided for "computerphobia." Invariably, our employees were intimidated by the technology at first, but, in a short period of time, interaction with the personality of Merlin won them over.

Students learned to solve problems on their own, and they learned to resist the temptation to ask someone for a solution. They began to appreciate their ability to control the learning process themselves instead of conforming to someone else's learning style. Most importantly, the majority of learners had fun interacting with the course material.

Following the initial field test, additional students were accommodated at the original locations with no additional effort from the main office staff. We believe there is no limit to the number of employees who can now learn about CIS with little additional expense to First Union.

Interactive video solved problems for us that we could not solve with traditional methods. We were impressed with the results of the program, and with favorable acceptance from our employees. Interactive video has turned out to be a pioneering effort that shows great potential.

HANDYMAN/HOMERS: IN-STORE TRAINING*

As a part of the Human Resource Department, the Handyman/Homers video group is currently staffed by three full-time professionals and serves 80 home centers located in six different states. Each home center has a system consisting of a Betamax video tape recorder (VTR) and a monitor. Previous video training programs have dealt with employee orientation, benefits and position training.

Use of Interactive Video

In the process of developing a video series for store training, it was determined that the fourth module, cashiering, would work well as an interactive program. The purpose of this training program was to create an awareness of the concepts of cash accountability, loss prevention and customer service. The cashiering training module was made interactive to keep interest levels high and also to guarantee as high a level of information retention as possible.

The video portion of the training program shows a cashier as he or she works through several typical transactions. With the help of voice-over narration, the trainee is able to hear many of the cashier's thoughts as he or she works through the day. While this training program does not teach the trainee to operate a cash register or process every type of transaction, it does demonstrate many of the company policies and procedures involved in the cashiering process. The trainee is scheduled to take this training program after being on the job for approximately one week.

Design and Program Features

Handyman/Homer's utilizes the Video-Dex Master Controller as its interactive video device, which is linked with a Sony Betamax SLP-300. The training is independent, self-paced and accompanied by a workbook. The workbook explains to the learner what the program is all about, how the training program works, how to use the program, how to use the Video-Dex and gives detailed explanations and follow-up information for each individual segment. At the end of each segment there are one to three interactive questions on the video tape. The learner is presented with the multiple choice questions in the form of character generated graphics and frame store visuals.

Company reaction has been receptive to interactive video, with the most

---

*Material for this case study was contributed by Tom Parks, of The Handyman of California, Inc. (San Diego, CA).

enthusiastic feedback coming from the users themselves. Users report that attention remains high during the training and that references to the training programs are made frequently after the learner returns to the job.

Interactive video training has proved to be cost effective time and again. Given the present economic conditions, this is a big positive. In addition, interactive video training frees management personnel to conduct their regular job duties and responsibilities while learners are being trained. Also, we can be assured that employees are receiving the same quality instruction at each location, each time training is given. In short, interactive video has allowed us to standardize our training company-wide, and guarantee its results.

JEPPESEN SANDERSON: FLIGHT TRAINING PROGRAM

by John G. Girod

For nearly two decades Jeppesen Sanderson, the Denver-based subsidiary of the Times Mirror Co., has pioneered in the application of audiovisual media and measurable training objectives to flight training. This effort has achieved a marked degree of success in student comprehension and retention.

As interactive video technology moved from the laboratory into the area of practical application, training specialists at Jeppesen Sanderson quickly recognized that this new technology was the next logical step in developing company training systems.

As vice president Vernon L. Francen put it, "For us interactive video offers two advantages. First, it enables us to offer both the student and the training organization the time efficiencies of individualized computer-assisted-instruction through the program branching capabilities of the interactive format. Second, the video capability allows us to add the impact of visual stimuli to increase student interest and motivation."

Since time efficiency and student motivation are two important criteria in the development of Jeppesen Sanderson's audiovisual training programs, interactive video is seen as enhancing those capabilities in a video-based training system.

Use of Interactive Video

How does a developer of training systems go about developing an interactive video training system? At Jeppesen Sanderson, the process ideally follows ten steps. (In actual practice, that number will vary as the process is adapted to a customer's specific situation.) The steps are as follows:

1) Meet with the customer to establish the kind and extent of training desired.

2) Complete a task analysis of the functions involved in the training situation.

3) Use the identified tasks and skills isolated in the analysis, combined with competency levels established by the customer, to establish the desired learning objectives for the training system. In most cases, these objectives will be achievable, observable and measurable.

4) Develop a flow chart using the desired learning outcomes to establish the most efficient learning path for the training system, providing for the necessary program branching at decision points.

5) Write the computer program.

6) Write the narrative script for the learning path and the program branches.

7) Shoot the video sequences established by the flow chart/script.

8) Prepare supplemental print material (if any).

9) Assemble the completed system.

10) Revise as necessary in response to student feedback.

While the entire process is interlocking and interdependent, two steps are central to the accomplishment of the total process. According to Dale R. Benson, project development editor, the two steps that establish clearly what is to be done and how it is to be accomplished are 1) defining the learning objectives, and 2) creating the flow chart.

Benson explains the importance of these two steps as follows:

> "This process is designed to teach through a building block approach. Major learning objectives are achieved by having the student successfully complete in sequence smaller, subordinate learning objectives. Therefore, for the system to be successful, the objectives must be carefully drawn to establish what will be learned. The flow chart is the blueprint of how to accomplish the task. The flow chart serves the programmer, the writer, the photographer, the artist and the editor as the specific master outline to follow in achieving the learning objectives."

Program and Design Features

How does the system work in actual practice? Tracing a project through the system may answer the question. (While the project described is hypothetical, the decisions and results reported are based on a compilation of several actual projects.)

For many years, Jeppesen Sanderson has produced audiovisual training systems for both aviation and industrial/commercial customers. When interactive video training systems became technologically feasible, training experts at Jeppesen Sanderson recognized an advantage offered them by the new technology: judgment training. For example, they could set up a particular flight situation, ask the student to decide on the action to take and actually show him the results of that decision. (Thanks to the electronic wizardry of video, some wrong decisions can lead to heart-stopping results . . . and excellent learning.)

Meetings were then held with a number of companies to explain the advantages now available through interactive video training. XYZ Aircraft Corp. agreed to purchase an interactive video training course.

Since this new course was essentially a conversion and expansion of a course Jeppesen Sanderson was presently producing for XYZ Corp., the basic research was still valid. A new task analysis was not needed. After review, the basic learning objectives remained the same. A few objectives were modified, and several were added to incorporate the proposed judgment training.

A flow chart was constructed using the existing passive audiovisual course as the basic learning path, but incorporating multiple branching at the decision points. Alternate solutions ("varying degrees of disaster" as one programmer called them) were developed to illustrate the difficulties that could follow from improper decisions.

Once the learning objectives were revised, and the flow chart complete, the revision job began in detail. Writers, photographers, artists—all created their part of the mosaic being assembled. The greatest workload fell upon the photo lab/video studio. These technicians had to completely reshoot the visual portion of the lesson to convert it to video tape, plus add the branching portions; rerecord the narrative, incorporating the revisions and the new material; edit the entire project; and finally replicate the new video training system. The task of this revision was simplified somewhat because Jeppesen Sanderson has a complete in-house writing/editing staff, production art department, still and motion picture photo lab and two video production studios with accompanying editing facilities.

Evaluation

The student response to interactive video training systems is generally favorable. One immediate student response with Jeppesen Sanderson interactive programs is that "when a person knows that the program is going to ask him to respond to a question or make a decision, whether the program is an actual lesson or only a demonstration tape, concentration on and attention to the material is observably increased."

Research done several years ago involving Jeppesen Sanderson audiovisual training materials indicated that increased student response, comprehension and retention resulted from the careful use of audiovisual materials in training situations. A similar response is anticipated from research involving interactive video materials.

Thus, for the producer of training materials in the commercial market, interactive video offers the opportunity for increased flexibility of presentation, increased participation of the student in the learning process and—as a result of such participation—increased student success in learning.

MASSACHUSETTS INSTITUTE OF TECHNOLOGY: SURROGATE TRAVEL*

by Robert Mohl, Ph.D.

Use of Interactive Video

The Interactive Movie Map is a computer controlled video disc-based system which simultaneously displays two representations of an environment—travel land and map land. Users are allowed to explore both representations interactively.

In travel land, the users "drive" around familiarizing themselves with the space by seeing real footage previously filmed on location. In this mode, called surrogate travel, they have complete control over speed, direction of travel, direction of view, route selection and even choose which season of the year they travel in. In addition, users can access individual building facades and see slide shows and facsimile data concerning each location.

In map land, users "helicopter" above aerial overviews of variable scale (and resolution) with customized navigational aids. Such aids include position pointers, route plotters and personalized landmarks. With the dynamic overview users are able to move their viewpoint above the surface (pan and zoom), change the mode of representation by flipping between map and photo, and access auxiliary spatial data.

*This paper was presented at the 1980 Midcon Professional Program.

The work reported herein has been supported by the Cybernetics Technology Division of the Defense Research Agency, under Contract No. MDA-903-78-C-0039. Project director: Andy Lippman. Principal investigator: Nicholas Negroponte.

Program and Design Features

The first step in constructing the Movie Map consisted of filming all the visual material to be accessed by the user. A small town, approximately 10 blocks by 15 blocks, was chosen as the site. The structural backbone of surrogate travel was the footage which was filmed driving down the center of every street, in both directions, and making every turn from each street onto each other intersecting street. The filming rig had four cameras pointed at right angles, shooting a single frame every 10 feet.

Pictures were taken separately of the 2000 facades in town, in both fall and winter, with careful registration between seasons. Wherever historical photographs of buildings existed, present day duplications were shot. In addition, slide shows were produced showing examples of cultural activity and facsimile data associated with hundreds of selected facades.

An aerial overview was constructed from two modes of representation: photo and map. The source for the former was a four by eight foot photograph assembled from smaller rectified aerial prints. The map was custom-designed for video legibility and to match the photograph. The size and location of map labels were carefully planned so that for each possible viewing window, the labels most appropriate for that level of scale would be present and readable. (For example, increasingly detailed information was displayed in smaller fonts which would be conspicuous only by moving the viewpoint close up.)

Implementation

The main elements of the Interactive Movie Map user station are two video display screens; one showing ground level surrogate travel and the other, mounted horizontally, showing the world map of the aerial overview. Both screens are touch sensitive and are typically the primary user interface.

Along the bottom of the travel screen is a menu control panel which the users touch to drive around. Functions include speed, direction, view and turn signals. The rest of the screen displays the continuous video disc footage called for by the user's decisions. When users want to access additional information at a particular location, they touch that area of interest. They are then presented with a front view of that facade (or open space), with the accompanying name superimposed over the picture and synthesized sound giving verbal information. At that point, users have the option of returning to travel or "entering" the facade to see an auxiliary slide sequence of cultural/facsimile data. In many cases, users have a third option of flipping the season selection from fall to winter, or the time control from 20th century to 19th century, to see an identically registered picture of the building in the mode chosen.

The map or aerial photo displayed on the overview screen serves primarily as a navigational aid. Another aid is a "you-are-here" pointer which is automatically updated as users drive around. A route plotter traces the most recent course of travel (with intensity fading away over time).

Personalized landmarks are overlayed on the overview in one of two ways. In the implicit mode, symbols are automatically placed at the locations where the users have accessed facades and slide shows during surrogate travel. In the explicit mode, users specify those landmarks which provide the most effective orientation cues. They make a quick "rotoscoped" sketch of the features pictured on the travel screen which are the most important subjective identifiers of a location. A small icon of this sketch is inserted at the appropriate place in the overview.

The overview display also serves as an active input interface for two functions. First, it can be used as autopilot control for ground level travel. Users touch beginning and destination points (with or without tracing the path in between) and surrogate travel "chauffeurs" them along the appropriate route.

Second, the overview serves as an interface to spatial data access. By helicoptering above a world view of expandable scale, users can zoom in on any square block to see more detailed information at increased resolution. This action is initiated by touching the target area on the screen. At any point users can flip the mode of representation between map and photograph. A detailed landmark map of building sites is also an option when zoomed close up. Just as in ground level travel, users can access pictures of facades and the accompanying cultural/facsimile data by touching the top of the building in the aerial view.

Equipment

The Interactive Movie Map employs several channels of communication with a host computer. User control is implemented through any of four interfaces which have equivalent functions: keyboard, joystick, speech recognizer and touch sensitive display. The computer generates graphics, produces synthesized sound and controls all commands to the video disc players. The video discs have a capacity of 27 minutes of real time play or equivalently 48,600 individually addressable frames. Any frame is randomly accessible with a maximum search time of about five seconds. The standard video disc commands include searching to a specified frame number, playing successive frames forward or backward at specified rates, freezing on a single frame, etc.

The video signals fed to the two display monitors, showing travel land and map land, originate from three optical video discs and two frame buffers. The three identical video discs are the storage source of all previously recorded imagery including travel footage, facades and auxiliary slide shows, maps and aerial photos. The frame buffers store run time generated graphics symbols including menus and navigational aids. All these signals are passed through a video switcher/mixer.

The switcher is used to alternate between video disc players. This technique allows search discontinuities to take place on the nonvisible disc before the new frame is cut to by the switcher. In the case of surrogate travel for example, the nonvisible disc always searches ahead to the next turn sequence, which can be cut to when the end of the block is reached. Alternating searches maintains visual continuity; staging ahead eliminates almost all delays. The nonvisible disc is shared between travel land and map land, serving whichever mode is active.

The video mixer is used to combine the computer-generated graphic overlays with the video disc image. The graphics overlayed on the surrogate travel screen include the menu control panel and text for the auxiliary facade data. The graphics overlayed on the overview screen include a mixed menu (similar to the travel screen), and the set of navigational aids relative to position and scale. Icons for landmarks, position pointers and route plotters are panned and zoomed isomorphically to the helicoptering of the video disc overview. Such real time movement is necessary for the overlayed symbols to appear correctly as stable features on the background.

Evaluation

The conceptual basis for the Interactive Movie Map is to provide the user with much of the experience of real world travel and personalized navigational references. A pilot study has been conducted to evaluate the kind of spatial learning that takes place with the surrogate travel portion of the system. A second study is currently being undertaken to look at the integrated system.

The pilot study showed that users find it easy and natural to drive around in an unfamiliar environment making real-time path-choosing decisions. The process allows them to construct their own internal representations or cognitive maps of the space. Moreover, these cognitive maps are in many ways similar to the maps developed by travelers learning in the real space. The most compelling evidence confirming the success of the system came from Movie Map subjects who subsequently visited the town site for the first time. They typically reported that they knew the town so well it was as if they had been there before.

It is clear that surrogate travel must ultimately fall short of the real world experience; the video display is not a window onto a live reality. Users did not always interpret the scale of the space accurately (heights of mountains and linear distances for example). Some of the shortcomings have been overcome since the study by refining the user control interface and by increasing smoothness and camera stability.

Conclusions

A number of techniques have been developed to optimize the use of optical video disc technology. The number of players necessary is taken to be at least one greater than the number being displayed at any one time. This combination allows all visual discontinuities (searching) and most delays to be hidden from the user by execution on the nonvisible player.

Another principle with the video disc is to maximize the variety of playing options that do not require searching. In surrogate travel for example, the left and right turns at the end of a block both begin "head-to-head" on the same frame number. One can be viewed by playing the disc forward, the other by playing back. In the future, this same principle may be used to expand the options when players are able to access every other (or third, etc.) frame, or opposite sides of the disc, during the vertical retrace interval. In this case parallel sequences could be interlaced.

The pilot study of the surrogate travel portion of the system alone has demonstrated the need for accompanying aerial overviews. In evaluating the integrated system, it is expected that duality of representation is one of the concepts that will prove to be important in the learning experience. This concept of seeing the same thing from different points of view was applied to many different scenes—ground/aerial, global/local, map/photo, fall/winter. The implications are likely to extend to more general video disc teaching applications which have a need for multiple levels or languages of explanation, multiple rates of time flow, multiple variations of an event, multiple formats of a problem.

A host of interactive video disc-aided instruction applications are on the horizon. One advantage of the Movie Map is that the script for branching through the network of paths is dictated by the topology of the space. Any subject matter that has inherent network structures, no matter how far removed from geographical space, can be formatted using similar principles for implicit authoring. Examples can be as varied as chemical structures, biological systems, political evolution, socioeconomic patterns. The consequence is that the video disc can become not just a vehicle for prescribed learning paths, but a reference instrument driven by the user's research interest.

REFERENCES

1. D.S Ciccone, B. Landee and G. Weltman, "Use of Computer Generated Movie Maps to Improve Tactical Map Performance," (Woodland Hills, CA: Perceptronics, Inc., Technical Report PTR-1033s-78-4).

2. W. Donelson, "Spatial Management of Information," SIGGRAPH '78 Proceedings, *Computer Graphics* 12 (3) (August 1978).

3. A. Lippman, "Movie Maps: An Application of the Optical Videodisc to Computer Graphics," SIGGRAPH '80 Conference Proceedings, *Computer Graphics* 14 (3) (July 1980).

4. R. Mohl, "Cognitive Space in a Virtual Environment," unpublished paper (Cambridge, MA: Massachusetts Institute of Technology, 1979).

U.S. ARMY SIGNAL CENTER: HANDS-ON TECHNICAL TRAINING

by William D. Ketner

The Army is confronted with an enormous technical training task, complicated by the training of large numbers of enlisted personnel on large quantities of complex equipment. The U.S. Army Signal Center at Fort Gordon, GA, is responsible for the Army's communications/electronics training. The Signal Corps has 54 military occupational specialties (MOS) and over 100 technical courses in support of these MOSs. These courses range in duration from 10 to 48 weeks. The curriculum sophistication can be as simple as learning to climb telephone poles and stringing wire, or as complex as repairing computers and satellite communications ground station equipment. The Signal School trains approximately 35,000 soldiers per year in residence.

Acquiring and maintaining some of the communications equipment is becoming a problem for the Signal School. For example, a problem was encountered in the Satellite Communication Ground Station Equipment Repair course. The course is 36 weeks long and provides enlisted personnel with the skills, techniques and knowledges necessary to operate, troubleshoot and repair the digital communications subsystem and satellite communications ground terminal. The specific problem was found in a three-hour practical exercise lesson for online test of the multiplexer. Certain switches are designed to be turned on and off for perhaps 100 times during the life expectancy of the multiplexer. Because the equipment is being used for training, the switch is being turned on and off 100 times a week. This causes extensive failures. The approximate cost of the multiplexer is $40,000, the one time switch repair cost can be as high as $1000 and the down time for the switch is often nine months, because it is not stocked as an off-the-shelf item.

Use of Interactive Video

The problem, then, was to provide the student with simulated hands-on training, independent of equipment, on the multiplexer, and ensure that the student can perform the task, thus avoiding excessive cost for equipment and repair. To solve this problem the Signal School developed a hands-on simulation of the online test of the multiplexer using the Apple II microcomputer, a DiscoVision Associates (DVA) video disc and a light pen for interaction. The typical cost of this "off-the-shelf" equipment is about $5000.

The video disc, microcomputer and light pen are integrated to form a student interactive system, which provides students the opportunity to learn and practice online testing of the multiplexer. The video disc can provide up to 54,000 frames (pictures of equipment and operator test functions) plus sound on each side. The microcomputer provides for an interaction with the video disc, so that individual frames or a series of frames can be accessed. The light pen is the tool the students use to interact with the system. If the student is required to make cable connections, he or she simply touches the light pen to the appropriate connector on the screen and a connection is made by the microcomputer accessing the correct frames on the video disc, which then displays the completed connection. If the student makes a mistake, the system will indicate that an error has been made and that the student should try again.

Program and Design Features

The procedures for developing the interactive video disc simulation system are quite involved. First, a demonstration video tape of all the online test functions is made. This tape, with numbered frames, is sent to the video disc manufacturer for mastering (transfer to disc). A computer program is developed that interacts with the TV screen and light pen, and drives the video disc to the appropriate frame or frames (the video disc player has real-time random access capability).

The system is set up in a demonstration mode, a simulation mode and a test mode. In the demonstration mode, all the operation functions for the online test are demonstrated by an instructor. This is used to teach the online test procedures and for remediation when required. The simulation mode is used for students to practice the online test procedures by interacting with the system through the use of the light pen. The test mode is used to examine the student and to record efficiency in performing the required task.

Evaluation

After the interactive video disc system was produced and debugged by the Educational Technology Division staff at Fort Gordon, an experimental analysis of the project was made. In November 1980, 51 students were scheduled to begin training. The students were randomly assigned to either the experimental group or the control group. The experimental group practiced the online test procedures using the interactive video disc system. The control group practiced the online test procedures using the actual equipment. The research report concludes the following: All tests supported the conclusion that no significant difference was found in the performance of the two groups.

In summary, the interactive video disc system has been used in the classroom for more than 10 months, allowing the students to practice the online test while final testing is given on the multiplexer. No hardware problems have developed, and only minor software problems have been encountered. The students and instructors like to use the system. In a small way, this has helped solve the equipment cost and equipment maintenance problem. The Signal School currently has several other interactive video disc systems in various stages of development. It is anticipated that this training concept for simulated hands-on practical experience can save the Army millions of dollars in training equipment and maintenance costs, and the soldier can still achieve the desired performance standards.

Selected Bibliography

Bennion, Junnius L. *Authoring Procedures for Interactive Videodisc Instructional Systems.* Provo, UT: Brigham Young University, Institute for Computer Uses in Education, 1976.

Cannon, Don L., and Luecke, Gerald. *Understanding Microprocessors.* Dallas, TX: Texas Instruments Learning Center, 1979.

Ciccone, D.S.; Landee, B.; and Weltman, G. "Use of Computer Generated Movie Maps to Improve Tactical Map Performance." Technical Report PTR-1033s-78-4. Woodland Hills, CA: Perceptronics, Inc, 1978.

Donelson, W. "Spatial Management of Information."SIGGRAPH '78 Proceedings. *Computer Graphics* 12 (August 1978).

Flemming, Malcolm, and Levie, W. Howard. *Instructional Methods Design.* Englewood Cliffs, NJ: Educational Technology Publications, 1978.

Gagne, R.M., and Briggs, L.J. *Principles of Instructional Design.* New York: Holt, Rinehart and Winston, 1979.

Gerbner, G. and Gross, L. "Living with Television: The Violence Profile." *Journal of Communication* 2 (1976): 173-199.

Greenblatt, Stanley. *Understanding Computers through Common Sense.* New York: Simon & Schuster, 1979.

Kobert, Don, and Bagnall, Jim. *The Universal Traveller.* Los Altos, CA: William Kaufmann, Inc., 1976.

Lee, Robert, and Misiorowski, Robert. *Script Models.* New York: Hastings House, 1978.

Lippman, A. "Movie Maps: An Application of the Optical Videodisc to Computer Graphics." SIGGRAPH '80 Conference Proceedings. *Computer Graphics* 14 (July 1980).

Mohl, R. "Cognitive Space in a Virtual Environment." Unpublished paper. Cambridge, MA: Massachusetts Institute of Technology, 1979.

Nierenberg, Gerard I. *The Art of Creative Thinking.* New York: Simon & Schuster, 1982.

Samuels, Mike, and Samuels, Nancy. *Seeing with the Mind's Eye.* New York: Random House, 1979.

Schubin, Mark. "The Future of Television Technology." In *The Video Age: Television Technology and Applications in the 1980s.* White Plains, NY: Knowledge Industry Publications, Inc., 1982.

Sigel, Efrem, et al. *Video Discs: The Technology, the Applications and the Future.* White Plains, NY: Knowledge Industry Publications, Inc., 1980.

Periodicals of Interest

AV Communications Review. Washington, DC: Association for Educational Communications and Technology (AECT), quarterly.

Cavri Systems, Ink. New Haven, CT: Cavri Systems, quarterly.

E-ITV. Danbury, CT: C.S. Tepfer Publishing Co., monthly.

Electronic Learning. New York: Scholastic, Inc., 8 issues/yr.

Performance Instruction. Washington, DC: National Society for Performance and Instruction, monthly.

Training and Development Journal. Madison, WI: American Society for Training and Development, monthly.

Training Magazine. Minneapolis, MN: Lakewood Publications, monthly.

Videodisc Design/Production Group News. Lincoln, NE: Nebraska ETV Network/University of Nebraska—Lincoln, monthly.

Videodisc News. Washington, DC: Videodisc Services, Inc., monthly.

Videography. New York: United Business Publications, Inc., monthly.

Videoplay Report. Danbury, CT: C.S. Tepfer Publishing Co., monthly.

Video Systems. Oberlin Park, KS: Intertec Publishing Corp., monthly.

Video User. White Plains, NY: Knowledge Industry Publications, Inc., monthly.

Index

ABOUT THE AUTHORS

Steve Floyd is manager of video development for the training department of Coca Cola in Atlanta, GA, where he is responsible for managing the development and production of video-based training packages. Prior to joining Coca Cola, he was a writer/producer for the Texas Instruments Learning Center, and the manager of creative productions for Organizational Media System, Inc. Mr. Floyd has written a number of articles on video communication and training design, and frequently speaks at professional conferences and seminars. He has written and produced a wide range of programs for a number of corporations, and has received awards for his video programs from the ITVA and from the U.S. Industrial Film Festival. Mr. Floyd holds a B.A. in economics and English, and an M.S. in instructional systems technology from Indiana University.

Beth Goodwin Floyd is an independent producer based in Atlanta, GA, servicing clients such as IBM, Coca Cola and the National Presbyterian Church. Previously, as project manager for Organizational Media Systems, Inc., she provided consulting services to a number of clients including Mobil Oil Corp., Philips Petroleum and Miller Brewing Co. In 1980, she was media coordinator for the GTE Southern Region headquarters, overseeing media production for the firm's national training packages, as well as providing technical and marketing support materials, and managing GTE's on-site teleconferencing services. Ms. Floyd has worked on a variety of media projects from corporate to commercial, and in 1981, won the ITVA Golden Reel of Merit. She received her B.F.A. in communications, *magna cum laude,* from Southern Methodist University.

Michael L. Schwarz is producer/writer/director at Northrop Worldwide Aircraft Services' Northrop Aircraft Division in Los Angeles, CA, where he is responsible for interactive video disc production. Previously, as a media designer for Northrop, he produced instructional video tapes and was responsible for the design and development of interactive video programming. Prior to joining Northrop, he worked in the TV studio at the Texas Instruments Learning Center, and was on the staff of the North Central Texas Regional Police Academy, where he helped develop and expand media production capability. Mr. Schwarz has conducted interactive video seminars at several national conferences, including those of the ITVA, NAVA and the AudioVisual Management Association. Presently, he is the ITVA regional vice president for Region X. He holds a B.F.A. in radio-television-film from Texas Christian University.

Kenneth G. O'Bryan is head of production and development at the Addiction Research Foundation in Toronto, Canada, where he is responsible for policy development, executive direction and general management of television and audiovisual productions, art and graphic design, and communications projects. Formerly director of formative research and special projects at the Ontario Educational Communications Authority, he has been an advisor to several organizations, including the Children's Television Workshop, the Corporation for Public Broadcasting, Warner Communications, the U.S. Federal Trade Commission, the Canadian Broadcasting Corp. and the British Broadcasting Corp. An expert in educational and industrial television research, design and production, Dr. O'Bryan has written numerous articles on education, applied psychology and educational communications management, and is author or co-author of more than 20 books on educational television management, production, writing and research. Dr. O'Bryan holds a Bachelor of Education, Master of Education and Ph.D. in educational psychology from the University of Alberta, Edmonton, and is a research associate at the Harvard University Center for Children's Media.

Patrick McEntee is president of ISO Communications, Inc. in New York, NY, a company that develops interactive video disc, video tape and videotext programs, and offers authoring support services including software and courseware development. Prior to founding ISO Communications, Inc. in 1981, Mr. McEntee held management positions with Sony Corp., ABC and Manhattan Cable Television. He has taught at a number of schools including the Columbia University Graduate School of Business, the Massachusetts Institute of Technology and New York University. His articles have appeared in a number of professional journals, including *Broadcast Engineering, Creative Computing, Videodisc News,* the *SMPTE Journal* and *Videodisc/Videotex.* In addition to receiving an IBM Fellowship in numerical control, Mr. McEntee was a term fellow at Oxford University, England, in 1972. He received his B.A. in computer graphics, with honors, from Yale College.

David Hon is the national training manager for the American Heart Association in Dallas, TX, where he has designed and implemented all manner of instructional programs for the staff and 2 million volunteers. He has been working in the field of interactive video for years, both in the design and development of interactive video hardware and software, and his current CPR Computer/Videodisc Learning System has won the NAVA award of excellence. Previously, he was a program manager at the Texas Instruments Learning Center. He is a two-time winner of the ITVA's Golden Reel of Excellence award. Mr. Hon has written numerous articles on interactive video, and is published regularly in national journals. In addition, he is a frequent speaker on the subject of interactive video at national and international conferences and seminars. Mr. Hon holds a B.A. in journalism and English from the University of Washington, and an M.A. in communications from the University of Tulsa.

Other Titles Available from the Video Bookshelf

The Home Video and Cable Yearbook, 1982-83
300 pages (approx.) softcover $85.00

The Video Age: Television Technology and Applications in the 1980s
264 pages hardcover $29.95

Practical Video: The Manager's Guide to Applications
by John A. Bunyan and N. Kyri Watson
203 pages softcover $17.95

The Video Register, 1982-83
300 pages (approx.) softcover $47.50

Video in Health
edited by L. George Van Son
234 pages hardcover $29.95

Video Discs: The Technology, the Applications and the Future
by Efrem Sigel, Mark Schubin, Paul F. Merrill, et al.
183 pages hardcover $29.95

Video in the 80s: Emerging Uses for Television in Business, Education, Medicine and Government
by Paula Dranov, Louise Moore and Adrienne Hickey
186 pages hardcover $34.95

Managing the Corporate Media Center
by Eugene Marlow
215 pages hardcover $24.95

Video User's Handbook, 2nd ed.
by Peter Utz
512 pages hardcover $24.95

Available from Knowledge Industry Publications, Inc., 701 Westchester Avenue, White Plains, NY 10604.